HOW TO BUILD A

BUGPROOF ROOM

HOW TO BUILD A

BUGPROOF ROOM

ANGUS GLAS

PALADIN PRESS
BOULDER, COLORADO

How to Build a Bugproof Room
by Angus Glas
Copyright © 1990 by Angus Glas

ISBN 0-87364-581-2
Printed in the United States of America

Published by Paladin Press, a division of
Paladin Enterprises, Inc., P.O. Box 1307,
Boulder, Colorado 80306, USA.
(303) 443-7250

Direct inquires and/or orders to the above address.

CONTENTS

CHAPTER ONE

MY GOOD FRIEND THE LAWYER

> A sekret ceases tew be a sekret if it iz once confided—it iz like a dollar bill, once broken, it iz never a dollar again.

> —"Affurisms" (1865) from *Josh Billings: His Sayings*, Henry Wheeler Shaw (1818-1885)

My good friend Robert is a criminal-defense lawyer. He specializes in defending persons who are charged with serious crimes. Most of his work is to assure that each client receives a *fair* trial, and he aggressively exploits the rights of his clients. Robert is well-known and successful. His law practice has always been in compliance with the ethical canons of the American Bar Association and the legal system in general.

Robert admits (privately) that many, and perhaps most, of his clients are guilty of some offense that is related to the crime with which they have been charged. He also claims that most of his clients have been entrapped or had evidence planted on them or in their cars or homes. Almost all of his clients have been charged with crimes that are more serious than the one(s) they actually committed.

Robert says the adversary system of justice in America encourages prosecutors to charge offenders with more serious crimes (which the prosecutors cannot prove) in order to intimidate them and force them to plead guilty (plea-bargain) to a lesser crime (which the prosecution may or may not be able to prove). By plea-bargaining to a lesser charge, criminal offenders avoid a trial and the risk of conviction on more serious charges.

Even though lawyers are aware of this ploy, they are obligated to explain the plea-bargain option to their clients, and they must agree to the plea bargain if their client, even under duress, tells them to accept the prosecution's offer. For example, if a client pleads guilty to possession of cocaine for personal use, he may get probation or a suspended sentence. If he goes to trial, he is vulnerable to fabricated evidence, perjured testimony, the zeal of a politically inspired or "end-justifies-the-means" prosecutor or law-enforcement agency—or even the whim of a distorted or vengeful jury. No one wants to wind up with a sentence of thirty years to life having been convicted of drug smuggling or distributing, exploitation of children, or other serious offense.

Because Robert is a successful criminal-defense attorney—he wins more than 85 percent of his cases—he has earned a place on the government's bad guy attorney list. Frankly, many politically inspired and crusading prosecutors and law enforcement officers would like to get Robert disbarred. Failing that, they would like to make his life so uncomfortable he'd get out of his criminal-defense practice and concentrate on noncriminal cases: business law, divorce, child custody, wills, or probate.

About a year ago Robert defended a man (I'll call him Mr. Washington) who had been arrested and charged as a drug smuggler. The assistant prosecuting attorney openly suggested that Mr. Washington was a big-time dealer they'd been "out to get" for some time. Several drug-enforcement agencies, working with the narcotics division of a major East Coast metropolitan police department, cooperated in the arrest of Mr. Washington. They handpicked Mr. Washington because they thought he was the weak link in a major cocaine distribution system that they knew was run by others, including the people who supplied Mr. Washington. The metropolitan district attorney, acting for the metropolitan police, charged Mr. Washington with importing and distributing cocaine even though their scant evidence only confirmed that he was a personal cocaine user who supported his habit by dealing to personal friends and fellow workers.

In reality, Mr. Washington was an ordinary, young, upwardly mobile professional, who casually flirted with cocaine. His eleven customers were all personal friends. Robert believes that Mr. Washington's primary incentive to sell cocaine to his friends and coworkers was his need for status—specifically, his need to be a big shot and the life of the party. Mr. Washington admits he has a strong desire to be liked and accepted and that he routinely goes out of his way to please and impress his friends and coworkers. At the time of his arrest, Mr. Washington was a junior accountant in a firm that employed more than 30 accountants. He thought of himself as the firm's token black employee. Mr. Washington was a college graduate, working to be a certified public accountant (CPA), who had no prior arrests. The inflated charges of drug smuggling and drug distribution placed a permanent cloud over his hope of becoming a CPA.

The metropolitan police, using the theory that Mr. Washington would tell his attorney, Robert, and others the truth about where he got the cocaine and to whom he sold it, obtained a court order to bug Mr. Washington's home telephone. An overzealous narcotics detective, seeing a chance to hang Robert and others on the Washington bug court order, misused the court order and installed the bug in the conference room of Robert's law office.

Obviously, Mr. Washington wasn't Robert's only client. Once the Washington bug was installed in Robert's conference room, the attorney-client privilege was effectively destroyed for all of Robert's clients. As a result, interagency cooperation between the metropolitan police unit that placed the bug and the narcotics units of several other agencies suddenly hit a new high. They discovered more than twenty new criminal relationships, the least important of which was Mr. Washington's low-level supplier and the list of casual-user friends and coworkers who were his eleven customers.

For a period of about two months, the courts approved a flurry of new telephone taps and room bugs against a wide range of persons who had not previously been suspected of anything. In each case, the probable cause offered as justification for the new court order was "an anonymous tip" or "information from a well-placed confidential informant."

Robert had been an assistant prosecuting attorney prior to entering private practice, so he quickly recognized that too many cases were being filed against individuals mentioned by his clients at attorney-client meetings held in the conference room of his law office.

Robert came to me because he thought his office was bugged. That same evening, he and I found the bug in his conference room. It was a simple but effective device, available to anyone for less than $100 from several sources, and which could be installed in ten seconds or less. The bug was a drop-in telephone pick-up, which replaced the original telephone pick-up in the mouthpiece end of the conference room telephone.

Professional-quality devices like the one we found can be activated from any telephone, anywhere in the world. The eavesdropper dials the number, then blows a dog whistle with a special pitch or uses an electronic tone generator to activate the bug before the bugged telephone has a chance to ring.

Because the bugged telephone does not ring, no one is alerted to the bug. Since the conversation is carried over the regular telephone line, there is no wiring or any radio signal to be detected. Once activated, the mouthpiece end of the existing telephone becomes an excellent microphone (especially if it is a speaker phone), and the telephone itself remains fully functional. To turn the bug off, the eavesdropper simply blows his whistle or uses his electronic tone generator again. Meanwhile, everything said inside the room can be heard and/or recorded from any other telephone.

If you value privacy and your civil rights, you'll be pleased to learn that Robert socked it to the errant narcotics detective in the end. Robert fabricated a criminal case, with the help of the supervisor of the internal affairs unit (IAU) of the same metropolitan police department that had illegally misused their court-authorized bug.

The reverse-sting worked like this: officers from internal affairs made a phony arrest of an IAU plant, who then came to Robert for his defense. Next, the plant and Robert discussed nonexistent criminal relationships between the plant and three prominent but innocent people who agreed to cooperate. Finally, in the conference room, the plant pretended to give Robert a retainer in cash, saying, "You know this is dope money. I hope it doesn't bother you that I got the money from the coke dealers in my distribution net?" Accepting dirty money is a no-no that can result in a lawyer's disbarment.

It took only twelve days for police to issue arrest warrants for Robert and the three cooperating individuals who had been identified by the internal affairs plant during the staged attorney-client conversations in the conference room. The police arrested Robert, but he was in jail less than forty-five minutes before the internal affairs supervisor arranged his release. The other three were never arrested. Internal affairs immediately quashed the arrest warrants for them.

Robert's reverse-sting operation resulted in the overturn of all of the convictions that were based in any way on information illegally gathered via the bugged telephone in his conference room, including the plea bargain accepted by Mr. Washington.

Right after Robert's reverse-sting operation I made some quick-fix modifications to the conference room.

It is interesting to note that the metropolitan police continued to consider themselves (as do many law-enforcement people) uniquely above the law. Robert tells me that, to the best of his knowledge, none of the offending officers was fired or suspended. He says that several federal drug-enforcement people transferred—to similar jobs—in other cities and that two metropolitan police officers rotated out of narcotics to other undercover assignments where, presumably, their bugging, tapping, and surveillance skills would again be put to use.

Unfortunately, reverse-sting operations are rare, and seldom are they as successful as the one done by my friend in cooperation with an honest and aggressive internal affairs unit.

★ ★ ★

Now that we've entered the law and order 1990s, telephone taps and room bugs are more and more common. Some are technically legal—as was the original, tightly restricted court-ordered bug intended for Mr. Washington's home telephone. But, as in Robert's case, many of these court-ordered bugs are misused. It is just too tempting to the law-enforcement officers who become calloused after years of installing and monitoring the usually tedious conversations obtained from room bugs and telephone taps. In practice, legal bugs and taps almost always generate accidental information. This found wealth of free facts and hot leads can't be attributed to the bug or even acknowledged as based upon bugged conversations—as in this case between Robert and his clients. As a result, information surfaces as anonymous tips or reliable information from a well-placed confidential informant.

Although bugging, tapping, and other forms of surveillance are pervasive, it is unlikely that you or I will be the subject of a court-ordered telephone tap or room bug. If we do come under surveillance by audio or optical devices, it is much more likely to be surveillance put in place by a competitor or some other enemy.

Our most serious risk comes from coincidental or accidental surveillance. At one time or another, each of us will talk to someone who is the subject of surveillance or bugging. If so, we get dealt in to whatever problem that person has because we become guilty by association with them. It is even more likely that we will come under surveillance because we're in the wrong place—a physical location that is under surveillance. We become vulnerable because what we think is a private or privileged conversation or transaction happens in a place or over a telephone (even a pay phone) that is under surveillance.

As a result, by coincidence and purely by chance, we, our conversation, and our transaction will be compromised. That chance surveillance or eavesdropping exposure makes us every bit as vulnerable as if we had been the intended target. And that vulnerability can lead to embarrassment, trouble, and frustration—for us and our friends, neighbors, and associates—whether we are innocent or not.

The Orwellian prediction of a "Big Brother" who watches over our private lives has become a reality—with a big boost from technical advances during the mid-1980s. By the year 2000, Big Brother, with the help of computer cross-indexing and universal police, will make personal freedom and independence impossible unless we learn how to protect ourselves from the invasion of our privacy now.

According to some social statisticians, 55 percent of all male Americans commit at least one felony before their thirty-fifth birthday. Prior to 1985, less than one-half of 1 percent were caught. By the year 2000 the number of persons caught—by accident or intent—is likely to skyrocket.

It may be helpful to think of safe conversation with the same degree of seriousness as, since the advent of AIDS, you now think of safe sex. Please read this book carefully. What you do to protect your privacy today will affect the rest of your life and the lives of your family, friends, associates, and all others who come in contact with you—just as they will affect your future.

CHAPTER TWO

MY FRIEND'S VULNERABLE CONFERENCE ROOM

Robert shares office space with two other attorneys. That evening, after we found the bugged telephone in his conference room, we also swept the rest of the offices.

The procedure was routine and simple enough. I placed a metronome—one of those tick-tock devices—in the conference room, then listened to an all-frequency, directional AM/FM scanner to see if the scanner picked up the regular beat from the metronome on any of the radio or television (AM, FM, VHF, UHF) frequencies it scanned. That only works, of course, when the bug you expect to find is sending a signal (transmitting).

I repeated the procedure several times over the following two weeks, each at a different time of day, always trying to activate any sleeping device with turn on probes (much like a hacker uses multiple signal sequences to open a garage door or access a restricted computer file). I was never able to locate any radio or television signals.

I also made a slow, methodical physical search for audio (listening) and optical (video) devices. I found none. I dismantled every telephone in each of the offices, then physically examined all of the light switches and electrical outlets, and each piece of electrical and electronic equipment, and every power source—in each case looking for multiplexing or hard-wired systems. Then I "patted down" every square inch of the surface area: walls, ceiling, and floor.

I routinely make a rough-scale drawing in grid form of each room I sweep. I allow about two hours for each 16' x 16' office. I also take apart every easy-to-assemble piece of office furniture—looking for hollow legs, false compartments in drawers, radio-activated transmitters in out-of-sight cavities, pinhole openings for through-the-wall "needlehead" microphones, camera lenses, or other intrusive devices. I personally turned upside-down every chair, table, desk, credenza, typewriter or computer stand, lamp, and all of the accouterments in the suite of offices that evening.

The following day, using powerful field glasses and my all-frequency directional scanner, I examined the exterior of Robert's law office (and the offices above, below, and to either side) from the nearest building across the street, looking and listening for an external retransmitter. Retransmitters can be driven by a feeble (low power) bug or a hard-wired crystal microphone from inside an office, through an external wall, for rebroadcast at much higher wattage to a receiver or call-forwarding device and/or recorder a mile or more away.

With the cooperation (for $50) of the night-maintenance supervisor, I also patted down the ceiling of the office below, the floor of the office above, and the common walls shared by offices on either side of Robert's suite.

Even though I searched carefully, the only audio surveillance device I was able to find was the original drop-in, tone-activated bug that had been placed by the metropolitan police narcotics unit in the mouthpiece end of the conference room telephone.

I was present when the supervisor of the internal affairs

unit arrived to examine the conference room telephone. He confirmed that the bug belonged to the metropolitan police narcotics unit and that it was the one authorized by the court order for installation in Mr. Washington's home phone. He told Robert and me that no other device had been authorized, and he acknowledged that the court order *apparently* had been abused. He agreed to leave the bug in place while he and Robert conducted a reverse sting—to be acted out by Robert and police officers assigned to his internal affairs unit.

THE PROBLEM FOR ROBERT

After my exhaustive electronic and physical sweep of Robert's law offices, we still couldn't be sure the place was free of listening and/or viewing devices—or that it would stay "clean." All that my careful search had done was raise the threshold of surveillance.

If Robert's conference room or office was bugged, we could assume that any audio or optical device would have to be:
1. an expensive, professional-quality, hard-to-get, subminiature model; and,
2. hard-wired so that no broadcast signal could be detected by an all-frequency scanner such as I used; or,
3. if it transmitted, the device would have to be:
    a. a remotely controlled "on-off" system that the eavesdropper could activate, as with the bug in the telephone; and,
    b. a very low-power transmitter that would require rebroadcast from a repeater such as I searched for from the building across the street.

All of these options are very expensive and more trouble than most folks who want information are willing to take. In other words, we escalated the threshold of how much time and money it would take to bug Robert, his fellow attorneys, and/or their clients during meetings in the conference room.

Although it is unlikely that anybody would go to that much expense and effort, we couldn't be absolutely sure that they would not. Nor could we maintain the integrity of the swept condition of Robert's office space without posting a trained, full-time guard (in whom we must vest the absolute and blind trust of whatever unexpected secret might be divulged in the clean room), or by repeating the entire sweep procedure immediately before every meeting at which Robert or his client(s) might be vulnerable should attorney-client confidentiality and privilege be compromised. Resweeping obviously precludes any spontaneous confidential conferences, which eliminated the value of the conference room for more than 80 percent of its regular use. Maintaining a guard was inappropriate because it was too expensive and didn't remove vulnerability.

There were, however, some relatively simple steps Robert and his fellow lawyers could take that would make it difficult to bug or visually invade the sanctity of their conference room.

I'll tell you what I installed for Robert in the next chapter. Following the procedures I took with Robert offers a practical solution to a problem faced by all lawyers who deal with criminal defense or criminal prosecution. The information may prove useful to any businessman, accountant, investor, or labor representative who is involved with business, taxes, leveraged buy-outs, acquisitions, takeovers, mergers, or any other high-value or high-vulnerability products or data that can range from new toys for the Christmas market to new fashions for the after-ski bunny market. And, of course, it would be useful to anyone involved in activities for which prison time and/or fines are a likelihood.

WHAT'S WRONG WITH THE CONFERENCE ROOM?

We will soon deal with each of the following and many other elements that are required to create a bugproof or bug-resistant room. But, quickly, here's what's wrong with

Robert's conference room:
- It has a window.
- It has a Muzak-type sound system.
- It has a telephone.
- It has one outside and three inside walls.
- It has heavy furniture and massive art displays.
- It has an impressive conference table.
- It has a floor.
- It has a ceiling.
- It has a door.
- It has no electrical grounding or "earth" on any wall, floor, or ceiling.
- It is visually vulnerable from the outside window and from the reception room.
- It has a single-entry funneled access through the reception room.
- It is open to casual use and anonymous visitors.

A Temporary Quick Fix for Robert

As a temporary measure, we eliminated the Muzak-type sound system and the telephone, and we escalated the level of difficulty for anyone who wanted to invade Robert's privacy through the window. As an interim step, we put a five-digit push-the-button solenoid activated lock on the door. Only the attorney partners were given the five-digit combination. Therefore, no more casual visitors could get into the conference room. Furthermore, no one was allowed inside the room unless he/she was accompanied by one of the partners who, personally, had to sign each guest in (even though it was agreed that guests could be signed in by code name).

First Fix: Get Rid of the Phone and Sound System

The first and most threatening problem to the law offices and the conference room was the telephone. It had already been bugged with a sophisticated professional device that took a narcotics agent less than ten seconds to 1) surreptitiously enter the room, 2) unscrew the telephone mouthpiece, and 3) replace the standard mouthpiece with the tone-activated "on-off" device. Obviously, we had to get rid of the instrument while, hopefully, turning it (like turning any other spy) to our advantage.

I *moved* the telephone. I didn't disconnect it. I did not want the physical instrument in the conference room, but I did want to know if anybody else was trying to play "let's turn on the phone bug." For the next sixty days, during the reverse sting, the telephone instrument "lived" and was appropriately answered—in a small storage room that continued to be used to store coffee filters, legal forms, letterhead, envelopes, and other supplies essential to any effective and modern office. I connected a voice or signal-activated recorder to the phone in the hope that anyone who tried to tone-activate the bug by dialing the conference-room phone number would have their command-to-activate tone sequence and their originating number recorded. I wanted to know *if* anybody was still trying to use the bug and, if so, *who* was trying to use it!

Secondly, the conference room, as did all other rooms in the suite of offices, had a Muzak-type piped-in background sound system. Although that is convenient and "white music" has an industrial (sometimes subliminal) value, it was a mechanism that put Robert, his fellow attorneys, and their clients at risk.

I personally disconnected the music system because, in order to play music, you must have a speaker. Every speaker is potentially a low-quality microphone when it is operated in reverse. That is, a microphone has a diaphragm, which is moved by any air disturbance (modulation) caused by noise, voices, or any other sound. The microphone converts these air disturbances or modulations into electromagnetic variations in voltage, which can be carried over wire and/or transmitted by amplitude (AM) or frequency modulation (FM/television).

In normal use, a speaker takes the same electromagnetic pulses and/or modulations and converts those pulses/modulations via a diaphragm into air disturbances (patterns) that we can hear and that we interpret as speech, music, or other noise. We most often hear them, with the help of electronic aids, through bass woofers or treble tweeters.

I physically removed the speakers and the wires that connected the conference room to the central music cable in the reception room.

Caller ID

Next, I had Robert subscribe to the telephone company's newly available Caller ID service. For less than $100, Robert had the telephone company install a small box with an on-line read-out to the telephone instrument previously located in the conference room. For a fee of less than $10 a month, the Caller ID will list the telephone number from which every call is placed, whether you answer your phone or not. As a result, Robert was able to record the calling number each time someone dialed the number of the former conference room telephone.

Second Quick Fix: Neutralize the Window

The second major problem with Robert's conference room—a problem shared by most business and professional conference rooms and executive offices—is that it is impressive. Impressive most often means that the conference room, along with the senior partner's office, is located in the most attractive and prestigious space that is available and/or affordable. Impressive also means that the furniture tends to be wood and heavy, the walls are decorated with massively framed prints and paintings, and the conference table is huge and substantial. Worst of all, impressive usually means that there is at least one window.

Windows make us all vulnerable for several reasons. Even though the window may be on an upper floor, we are at risk because anyone inside the room can be seen through the window. We and our associates are vulnerable to guilt by association. None of us wants to be seen with the wrong person or in the wrong place or at the wrong time. Each of us has our little secrets, and we are all vulnerable to accidental exposure.

Like it or not, there are lots of snoopy people who get their "innocent" kicks by peeking into other offices. (Remember the character Jimmy Stewart played in *Rear Window*.) Sometimes, just being seen with an individual causes gossip, and that gossip, heard by the wrong person(s), can impact our other business and social relationships. Furthermore, there really are people who can read lips—whether in person or from videotapes created by cameras equipped with long-range or low-light-level, light-amplification telephoto lenses.

There is, of course, another major window problem. Almost all windows lack resistance to radio signals, or in other words, they fail to dampen, squelch, or inhibit radiation from radio signals. As a result, even a very low-power FM transmitter—one that doesn't have enough power to penetrate a Sheetrock partition wall between rooms in a suite of offices—will carry a signal out through the window, high above noisy traffic, to a receiver or retransmitter across an eight-lane street.

Windows also make a room vulnerable to a relatively new audio surveillance device: laser reflection-refraction. When the laser was first developed, private industry (and the government) developed laser-beam listening (surveillance) devices. (It may help to think of a laser beam as a telephone wire.) Since the late 1980s, the price for laser devices has come down to the point where they are within the budget of most professional snoopers.

Note: Currently available laser devices can be operated successfully without the need to enter the room targeted for bugging.

To oversimplify the technology: a laser beam is focused on the inside windowpane from any line-of-sight location up to three miles away. Although clarity of signal and audio fidelity is enhanced if a tiny reflective dot is placed near the center of the inside windowpane, entry into the target room is no longer essential to basic eavesdropping. Flaws in the glass or acrylic pane, scratches, dirt, and even the pigment mass in acrylic coloring will serve as adequate laser-beam reflecting surfaces. When in use, the laser light stream is interrupted as conversation (as well as other sounds or noise) causes the windowpane to move slightly and thus deflect or refract the reflected laser beam. This minute movement of the windowpane causes changes in the reflected laser beam in exactly the same way a voice causes movement in the diaphragm of the mouthpiece of a telephone or microphone.

Like the telephone diaphragm, the variations in the reflected laser beam are amplified and converted electromagnetically on the receiving end into sound. More sophisticated laser listening devices use computers to enhance and filter out ambient noises such as wind, rain, or street or airport noise. Laser devices and all other listening systems use a process that is much like the familiar home stereos that amplify and convert electromagnetic pulses and amplitude modulations, and then output them through woofers and tweeters into what we perceive as sound. It may help to think of the basic laser signal variations as the minute signals on an audio cassette that are amplified and then translated into sound(s) through a car or home stereo system.

★　★　★

In the following chapter you will see how I escalated the threshold of security at Robert's law office and how similar systems can make it expensive, difficult, and time consuming to bug *any* office.

It is essential to recognize that while it is possible to make a bugproof room—and this book *will* document how to do it—it is often more practical (unless you or your client is paranoid) to make a room "bug resistant."

To adequately protect yourself and/or your clients, you

need only to assess and deal with actual risks. That is, you need only to estimate how much time, expense, and difficulty is likely to be used against you or your clients, and then establish a threshold that is more difficult and expensive.

A reality scale or test is appropriate. A bug-*resistant* room is usually adequate. For example, it is reasonable to expect that nobody will risk life or limb to learn if Sally is sleeping with Harry unless: there is one hell-of-a-big estate to be settled; one (or both) are religious zealots who believe they will go to some special hell if they don't get custody of the kids and rear them within their unique belief system; or someone has a compulsion to punish the "sinning" party or is a "nut case" for some other reason.

Only a bug*proof* room will protect your dialogue from the zealous nut, whether he or she is a crusading cop, prosecutor, estranged spouse, paranoid drug dealer, or wronged client or associate whose innocence or machismo has been violated.

Let's deal next with the how-to details of the quick-fix for the suite of law offices occupied by Robert and his associates.

CHAPTER THREE

A QUICK-FIX FOR WINDOW LEAKS

BASIC BUILDING BLOCKS, OR
HOW TO MAKE A TRIANGULAR SANDWICH

R. Buckminster Fuller created a geometric device that enriched all of us—the geodesic dome. Fuller showed how to use a basic triangle to form a curved dome. Join enough of Fuller's "triangle blocks" and, presto, you construct a house or wall or cover a football field with a lightweight, economical, sturdy dome. In theory, Fuller's geodesic dome could cover an entire city.

Our interest in Fuller's fascinating discovery extends only to his popular system for making triangular sandwiches. Unless we want to get fancy, we don't have to concern ourselves with the more complex problems of dome curvature or the three-dimensional angles of geodesic joints. Our triangular sandwiches will be used to construct flat, rectangular, soundproof panels. We will use our panels to

19

soundproof windows, doors, walls, ceilings, and even floors and other weight-bearing surfaces. Furthermore, since our finished surface will be flat and our triangular sandwiches separated by space, we don't even have to make our sandwiches perfect.

Sound Doesn't Travel in a Vacuum

Picture, if you will, a high-school physics class. The teacher places an alarm clock on a smooth stone slab. He winds the clock and sets the alarm to go off in two minutes.

The teacher then rubs a little grease on the open (base) edge of a science lab "bell jar." The bell jar is equipped with a small vacuum hose. The teacher then places the bell jar over the alarm clock and pumps the air out of the jar. The grease he has put on the base edge of the bell jar makes the interface between the jar and the smooth stone slab airtight.

In about thirty seconds, the little vacuum pump is laboring, and we know that nearly all of the air inside the jar has been exhausted. We have a partial vacuum. The teacher closes the valve on the vacuum hose to seal the jar and then turns off the vacuum pump.

About now we see that the alarm clock is "going off." Even though we can see the hammer strike the alarm clock bell and the alarm key unwind, *we can't hear the alarm* because there is not enough air inside the bell jar to carry the sound from the alarm clock to the glass side of the bell jar.

Although we can survive temporarily in a bugproof or sealed room, we cannot survive in an actual vacuum. That would be like an astronaut stepping into the vacuum of space without his space suit.

We can, however, separate ourselves from sound-sensing listening devices and human ears with a vacuum wall or a vacuum window. By itself, the vacuum wall or window won't protect us from electronic surveillance, but it will solve the problem of audio surveillance. By adding an optical blocking device and an electronic blocking device to our soundproof panels, we can keep out the visual and electronic snoops as well.

Note: The quickest, cheapest, and safest way to solve a window problem, of course, is to board up the window with ¼-inch grounded copper sheet metal, sandwiched between two soundproof acrylic panels. That blocks out the light, changes the outside appearance of the window, and is cosmetically offensive. Robert and his associates wanted to keep their window, not telegraph the modification to outside viewers, and maintain the appearance and decorum of the conference room.

Here's how we did it. For this example we'll use the specific dimensions of the single window in Robert's conference room. (See Fig. 1)

FIGURE 1
WINDOW DIMENSIONS

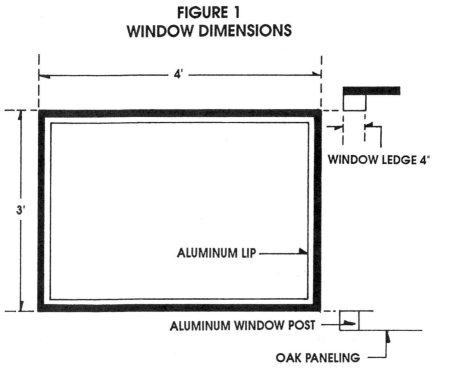

The window, which did not open, was 3′ high and 4′ wide. The lip of the aluminum frame into which the window was set covered 2″of the opening. The aluminum frame itself was flush with the outside of the building. As a result, it was recessed into a window box with an 8″ ledge, all of which was inset into the exterior wall of the conference room. All of the walls of the conference room were finished in oak paneling.

Because we needed visual and electronic blocking devices, we decided to make a triple-vacuum sandwich. To achieve this, we constructed two acrylic rectangular soundproof panels and one acrylic and mesh "spacer" surface, each with an outside measurement of 3′ 11″ by 2′ 11″—or an inch smaller than the inside dimensions of the actual window opening.

Tools and Supplies

The following tools are needed to construct a surveillance-proof window insert like the one I built for Robert's conference room.

One 4′ x 5′ (or larger) workbench or table.

One clean blanket, sheet, or other soft fabric.

One "fine-tooth" saw (either electric or hand).

One straightedge or metal carpenter's angle and marker.

One plywood pattern of a 9-inch equilateral triangle.

One hot-glue gun and clear (transparent) glue sticks.

One caulking gun and three tubes of silicone gel for the gun.

One electric drill.

One ¼″ drill bit.

One ⅛″ drill bit.

One soldering gun and solder.

One screwdriver.

One vacuum pump.

Note: Additional information about valves, vacuum pumps, hot-glue guns, caulking guns, decals, and other tools and supplies will be presented at the end of this chapter.

The following supplies will be needed:
9 each ¹⁄₁₆″ x 3′11″ x 2′11″ clear acrylic sheets.
4 each ⅛″ x 4′3″ clear acrylic sheets.
45 feet of ½″ or ¾″ wide polyurethane refrigerator door gasket material.
2 each air valves with shut-off locks.
2 each 3′11″ x 2′11″ fine mesh copper screens.
3 feet of bare copper grounding wire.
4 each ⅛″ x 1″ copper sheet metal screws.
1 single 3′1″ x 2′11″ decal.
50 each hollow cocktail straws.
2 each ½- to ¾-inch thick artificial kitchen "sponges."

To build the antisurveillance window package for Robert's conference room, I purchased nine sheets of ¹⁄₁₆″ clear acrylic precut into 3′11″ x 2′11″ rectangles. I later cut four of these sheets (using the plywood pattern as a tool) into triangles to be used as the "skins" of the individual equilateral (9″ x 9″ x 9″) triangles. Four of them would be used full size as the outer skins for the finished 3′11″ x 2′11″ soundproof panels. The ninth panel would be used to create the electronic blocking device.

The first task is to make a small plywood tool pattern measuring 9″ x 9″ x 9″. Since this pattern will be used to outline the 9″ x 9″ x 9″ clear acrylic skins, it should be sanded smooth so that it will not scratch the acrylic surfaces. Although I made triangular sandwiches that were 9″ x 9″ x 9″, it doesn't make much difference what size triangles you use to assemble the finished soundproof panels. I like to use smaller triangles if the finished panel is to be used as a floor or other weight-bearing surface and larger triangles if it is to be used for wall or ceiling areas.

To assemble equilateral triangles into a rectangular shape, you must also make four right angle triangles. I made them by dividing two of the 9″ x 9″ x 9″ triangles in half to make four right angle triangles approximately 9 ″ x 4½″ x 7⅜″. Since the triangles do not abut and are not individually airtight, it's not essential that each one be perfect. Construction of individual triangular sandwiches is necessary to maintain the shape and prevent the collapse of the outer panel when the air is exhausted from the finished rectangular panel.

Even if you have no carpentry skills, construction is simple if you make each angle of the equilateral triangles either 60 degrees or 120 degrees from the base of the triangle. Angles of the right angle triangles to be used in the four corners of your rectangle should be 90, 120, and 150 degrees. If in doubt, remember that there are 360 degrees in a circle and all other two-dimensional shapes: rectangles, squares, triangles, and so forth.

I like to space triangular sandwiches at least 1 inch, but not more than 4 inches apart, depending on whether the soundproof panel is for a wall or ceiling or for a weight-bearing surface.

To make an individual 9″ x 9″ x 9″ triangular sandwich, cut three pieces of clear ⅛-inch acrylic into boards measuring 1½″ x 9″. Cut three more pieces of ⅛-inch clear acrylic into boards measuring 1½″ x 8¾″.

Using transparent glue and a hot-glue gun, glue the flat surface of the three 9″ acrylic boards to the three 8¾″ acrylic boards to form three combined acrylic boards that are ¼″ x 1½″ x 9″ overall. One end of the ¼″ x 1½″ x 9″ board should be flush and the other end should have a ¼″ "step." (Do *not* cut the "step end" on a 45 degree angle for a "better fit." The gap created by the step will serve as a reservoir to be filled by hot glue and will help form a stronger bond.)

When the glue has set, form the three ¼″ x 9″ boards into a triangle, gluing the step end of each board to the flush end of the adjoining acrylic board to form a 9″ x 9″ x 9″ equilateral frame. Drill a ¼″ hole in the center of each leg of the assembled frame. (The holes may be drilled before assembly if you prefer.) The purpose of the ¼-inch holes and the space between the triangular sandwiches is to assure that all of the air inside the panel is exhausted when the vacuum is created. (See Figs. 2 and 3.)

FIGURE 2
TRIANGULAR SANDWICH

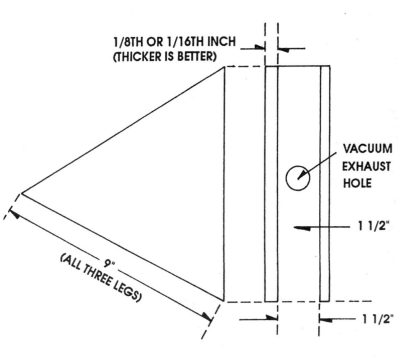

1/8TH OR 1/16TH INCH
(THICKER IS BETTER)

VACUUM
EXHAUST
HOLE

1 1/2"

1 1/2"

(ALL THREE LEGS)   9"

Next, cut two ¹⁄₁₆-inch clear acrylic skins into 9″ x 9″ x 9″ triangles. When the glue used to assemble the acrylic boards that were used to build the frame has dried, glue the ¹⁄₁₆″ x 9″ x 9″ x 9″ skin onto the frame. Be sure that no debris from drilling the ¼″ holes in the frame walls is trapped inside the assembled triangular sandwich.

If you wish, you may sand off any of the ¹⁄₁₆″ skin of the triangular sandwich that extends beyond the frame edge. *Do not sand the surface area of the skin* since abrasions on the surface of the skin can be used as a focal point for laser reflection which, since the laser is optical, will not be inhibited by the vacuum that will be created inside the assembled (rectangular) soundproof panel. Store the assem-

FIGURE 3
RECTANGULAR SANDWICH

1. Rectangular or square sandwiches can be used but are not as structurally strong as triangular sandwiches.
2. All sandwiches that abut a corner or wall must be bonded to the corner or wall and to the bottom skin.
3. Bond inside sandwiches only to the bottom skin. Arrange the inside sandwiches for best fit.

bled triangles on edge to prevent abrasions to the skin during construction.

To build a rectangular panel frame, cut four clear acrylic boards from the ⅛-inch stock material into 1⅝″ x 3′10⅝″ x 2′10⅝″ lengths and four acrylic boards 1⅝″ x 3′10¾″ x 2′10¾″. (The outside dimension of the panel frame should be approximately ¼-inch smaller than the outside dimension of the panel skins.) Using clear glue sticks in your hot-glue gun, attach the flat surfaces of the four shorter acrylic frame boards to the four longer frame boards so one end is flush and the other end has a ¼-inch lip, just as you did with the shorter (9″) acrylic frame boards for the triangular sandwiches.

Next, join the four ¼-inch frame boards you just assembled into a rectangle with an outside dimension of 3′10¾″ x 2′10¾″. Set the window panel frame aside for now.

Place a blanket or other clean, soft cloth on your workbench or table. Place the outside panel skin (3′11″ x 2′11″ x ¹⁄₁₆″) on the protective blanket.

Apply the optical screening device you have selected to the inside of one of the rectangular panel skins. A wide variety of screening devices can be used. For the optical portion of the blocking device in Robert's conference room, I used a large window decal I purchased from a hobby shop that specialized in mock leaded window supplies. When in place, the decal made the window in the conference room look like colorful leaded glass. Similar decals are available from auto specialty shops that sell privacy screens for van windows and from some hardware and building supply stores that sell similar screens for home picture windows, sliding doors, or other openings.

The purpose of the decal is to make it impossible for people to see into the room through the window and to distort any snooping laser probe to the extent that any reflected laser beam will be scrambled and unreadable. (The electronic blocking device is placed elsewhere in the window defense system and will be assembled later.)

Carefully remove the rectangular skin with the decal and stand it on edge out of the way. Now place the second rectangular skin on the cloth-covered workbench or table. Place the panel frame onto the clear acrylic skin. Hot-glue the frame onto the skin. It is important that the bead of clear hot glue be as generous and uniform as possible.

Arrange the four corners (the four right triangles) so they fit against the inside corners of the frame. Hot-glue the corner right-triangle sandwiches *to both the frame and the skin.* Next, arrange the other 9″ x 9″ x 9″ triangular sandwiches to your satisfaction on top of the skin inside the frame. Those triangular sandwiches next to the frame should touch the frame. Glue these outside triangular sandwiches *to both the frame and the skin*. Arrange the remaining triangular sandwiches so the space between them is relatively uniform. The inside sandwiches *should not touch each other*. Hot-glue them to the skin.

Note: The outside and corner triangular sandwiches that are glued to both the frame and skin will reinforce the frame when the vacuum is formed during final assembly. None of the triangular sandwiches will be glued to the decal. This ability to "creep" will allow a slight movement when the vacuum compresses and the pressure slightly distorts the outside skin on which you have applied the decal.

Use your caulking gun and silicone gel on the *outside* of the frame to seal the frame to the panel skin. The silicone gel will form an airproof seal that will not harden or become brittle. Since the atmospheric pressure will try to enter the vacuum, the heavy bead of silicone gel on the outside will be drawn into any gaps in the glue that bonds the panel frame to the panel skin.

Select the air valve of your choice and drill a hole through the inside panel skin (the skin *without* the decal) into which the valve will be seated. (Place the valve about two inches from the corner of the panel to allow space to install the panel into the window frame and to avoid the risk of the valve itself being used as a laser reflector.)

A large selection of air valves is available from gas station supply stores, school athletic and scientific suppliers, and from hobby shops. (I prefer propane hoses and valves.) Your choice will depend on the type of vacuum pump you use to exhaust the air from the finished panel. Seat the valve using appropriate bonding material. You may want to reinforce the seal with a bead of silicone gel *on the outside* of the assembled soundproof panel. Be sure that no debris from the valve hole remains inside the panel before you assemble it. (Debris trapped inside the panel or inside the triangular sandwiches will be sucked into the vacuum hose and may damage the valve or vacuum pump.)

When the valve is bonded, glue the skin to which you have applied the decal (with the decal on the inside) to the outside frame of the rectangular panel now on your workbench or table. (Do *not* glue the triangular sandwiches inside the frame to the decal.) Use the caulking gun and silicone gel to seal the skin to the frame as before.

Test the airtight quality of your glue and silicone seal by forcing smoke through the valve into the assembled panel. If any smoke escapes from the panel, reseal the flawed area with more hot glue and/or silicone gel. When you are satisfied that the finished panel is airtight, exhaust the smoke and residual air through the valve with the vacuum pump.

Close the valve before you turn off the vacuum pump, then disconnect the pump.

The outside soundproof panel is now complete and ready to be installed next to the building's original window.

Place a strip of ½″ or ¾″ wide polyurethane refrigerator door gasket on the aluminum window frame. Most refrigerator door gasket material comes in rolls and has adhesive backing. When the polyurethane gasket has been installed on all four sides of the aluminum window frame, put a continuous bead of hot glue on the gasket and quickly press the finished soundproof rectangular panel in place.

You may wish to use small shims on the bottom or sides of the panel to hold it in place. You may also wish to screw keepers into the side ledges of the window frame. *Do not screw anything into the soundproof panel.* Be careful not to destroy the integrity of the vacuum inside the soundproof rectangular panel which, now that it is installed, should fill the window cavity next to the original window. There should be about ½″ of free space on all sides of the soundproof panel, which is separated from the aluminum window frame only by the polyurethane refrigerator door gasket to which it is glued.

When you are satisfied that the soundproof panel with the optical blocking device you have just installed is properly seated and secure, apply the ½" or ¾" wide polyurethane refrigerator door gasket material along the border of the installed panel in the same way you previously applied the gasket to the aluminum window frame.

The Electronic Blocking Device

Hot-glue one of the fine mesh copper screens to one of the 2'11" x 3'11" clear acrylic skins. Solder about 5" of bare copper wire to the bottom left and upper right corners of the copper mesh.

Turn the acrylic skin over and hot-glue the second fine mesh copper screen to the second side. Solder about 5 inches of bare copper wire to the corners of the copper screen not previously used on the first side of the skin.

Drill four ⅛" holes in the metal frame of the window ledge about 1½ inches from the recently installed polyurethane gasket. Apply a continuous bead of hot glue to the exposed surface of the polyurethane gasket and quickly fit the electronic blocking device (the skin on which you have mounted the fine copper mesh) in place against the polyurethane gasket. Make sure all four wires are on *your side* of the device—inside the room with you—when the device is in place. (Two of the wires will have to be brought around the outside edge of the skin. The polyurethane gasket will easily compress to allow this, and no notch needs to be cut into the skin or wire mesh.)

Attach the four bare copper wires to the copper sheet metal screws and screw them into the grounding holes you previously drilled in the metal window ledge. Solder the wire to the screw heads once the screws are in place.

The Second Soundproof Panel

Now that you are an experienced builder of soundproof panels, the second will be a snap. Use the same materials and tools and the same procedure as before. You will not need to install the optical blocking device, the decal. Be

FIGURE 4
ELECTRONIC BLOCKING DEVICE

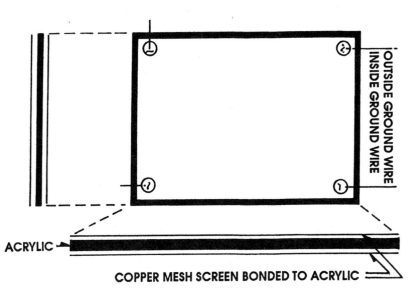

sure that the triangular sandwiches next to the rectangular frame are glued to both the frame and the skin and that they and the inside triangles are *not* glued to the second skin.

When you have constructed the second soundproof rectangular panel, test it for airtightness with smoke, then use the vacuum pump to exhaust the air. Stand the finished panel on edge, out of the way, where the clear acrylic skin will not be scarred.

Cocktail Straws and Sponge Separators

Before you install the second soundproof panel, you must separate it from the electronic blocking device that you previously installed against the polyurethane gasket mounted on the first soundproof panel.

Cut fifty transparent hollow straws (available from most cocktail waitresses for a modest tip) to make one hundred straws each about 3 inches long. Use rubber bands to tightly wrap both ends of the cut straws into bundles of about twenty-five each. You will discover that the bundled straws, because they are round and hollow, are remarkably strong and make excellent spacers with which to separate the soundproof panel that you are about to install, from the electronic blocking device that is already in place. (Only corner spacers will be needed. Since no vacuum will be formed around the electronic blocking device, no center spacers or triangular sandwiches are required.)

Cut eight circles from the artificial kitchen sponge slightly larger in diameter than the bundled cocktail straws. Cut off any "scrub" type abrasive backing from the artificial sponge circles. Hot-glue the sponge circles to each end of the four sets of bundled straws.

Once the glue that holds the sponge circles to each end of the bundled straws has cured, hot-glue one straw-and-sponge spacer near each of the four corners of the electronic blocking device.

When these are in place, apply hot glue to the exposed sponge ends of the four spacers and install the second soundproof panel. Again use shims and keepers, as necessary, to support and hold the second panel in place. You may wish to use silicone gel to help retain this second panel.

Please note that clear or transparent material (except for the screening devices) has been used throughout the construction of this system. Transparent material is essential so you can visually inspect all nonblocking elements of the device after it is installed. You or your client should visually inspect the antisurveillance system before each use to be sure no one has tampered with it.

★ ★ ★

When completed, your window—like the one I installed in Robert's conference room—will:

1. Prevent the transmission of sound through two sets of vacuum panels;
2. Effectively block all but the most sophisticated computer-enhanced optical viewing and laser-operated audio device through the decal; and,
3. Effectively block electronic radiation (radio signals) through the two earth-grounded, fine-mesh copper window screens. (The twin, grounded-copper-mesh screens will capture all but very high-powered, easy-to-detect transmitters.)

NOTES ON SOURCES OF MATERIAL

Nothing required to construct soundproof panels or blocking devices is expensive or difficult to find.

Vacuum pumps can be homemade and cheap, or they can be expensive professional models. Fig. 3 shows a professional quality vacuum pump that costs about $650. It creates a very high vacuum, does it quickly, and operates silently. (Since individual panels are easy to handle, you may want to take the finished panels to the high-vacuum professional pump or rent a professional pump for a day.) You can also make a workable vacuum pump using duct-tape and an industrial strength or shop vacuum cleaner. Or you can buy a hobby shop model of a vacuum pump for under $30. The professional model is best. The less expensive models will work for most applications.

Note: Two "soundproof" panels will do as well with a low-level of vacuum as a single panel with a high vacuum.

Air valves are avilable from a wide range of sources. The valve selection you make will depend on the type of vacuum pump you use. (Match the valve connection to the connector on the vacuum pump.) Service stations and construction contractors use a lot of air hoses. However, most "quick connect" valves operate one way only. It may be difficult to install them in your wall. Likewise, athletic departments pump up a lot of basketballs, footballs, and other sporting equipment. Any valve that will hold the pressure needed to inflate an object should also resist the atmospheric pressure (about 15 psi) that nature wants to put into the vacuum (void) created by your pump. Connecting your vacuum pump to the panel is not enough. You must have an independent turn-off—a valve mounted on the panel or on a short piece of hose to which the pump is attached.

I prefer propane hoses and valves. They are built to contain explosive gas under pressure. Most stores that carry camping or outdoor cooking supplies sell propane barbecue grills, heaters, and other appliances. They also carry the propane hoses and valves used to connect external tanks to stoves. You want the valve and a short length of hose.

Hot-glue guns and clear glue sticks are available at most hardware stores. Convenient hot-glue gun kits, including about a dozen glue sticks, cost less than $15. You can buy packages of 100 clear glue sticks for $10 or less, which is more than enough to construct any bugproof room.

Caulking guns and silicone gel tubes are also available at most hardware stores. The caulking guns cost less than $10. Tubes of silicone gel that fit into the caulking gun cost less than $5. Three of them are more than enough to construct a sophisticated bugproof room.

Fine-tooth saws are available everywhere. The finer the teeth in the saw—either electric or hand—the smoother the edge of the acrylic sheets you cut.

Acrylic sheets are available at most building materials outlets and at larger hardware stores. The thicker the acrylic sheet, the more expensive it is. For example, a half-inch thick 4 x 8 foot sheet can cost $100 or more. Thin sheets, as called for here, cost as little as $10 for a 4' x 8' sheet. Bonding two thin sheets makes them stronger than a single sheet of the same thickness, and it costs only a fraction as much. In addition, offsetting the sheets to make "lips," as suggested in the instructions, adds corner strength to the assembled frames.

Decals and other optical blocking devices can be found

at hobby shops, auto paint shops, or some building supply outlets. There is nothing wrong with improvising a paper decal—or even a colored poster—since the decal is inside the vacuum area of the exterior soundproof panel.

Polyurethane gaskets are stocked by most hardware and appliance repair shops to replace the worn gaskets on refrigerator and freezer doors. The gaskets compress when the refrigerator or freezer door is closed. The job of the gasket is to keep the cold in and the heat out. Since for our purposes, they are used only as spacers, we don't care if they make an airtight seal or not.

Hardware, including copper screws and wire and soldering guns, is available everywhere and costs very little. If bare copper wire is not available in the electrical section of your hardware store, buy insulated *copper* wire and strip off the insulation. If you don't already have a soldering gun, they are available in most hardware and hobby stores for less than $15, including more solder than you will need.

ASSEMBLY OPTIONS AND TIPS

How you install soundproof panels and optical or electronic blocking devices is determined by your purpose and the unique conditions you must resolve. It is sometimes appropriate to modify the basic design in some way, such as:

Sealing the existing window. You can do this by caulking corners, seams, and joints with silicone gel and/or painting porous surfaces with rubberized paint.

Sealing the outside panel to the window ledge. You can do this by packing any gaps between the panel and the sides, top, and bottom window ledges with rubberized material and sealing it with silicone gel.

Clouding a panel. You may wish to deal with optical blocking by "inflating" the outside airtight panel with dense smoke—even colored smoke. If "inflated," a panel will block optical surveillance but it *will not* prevent sound transfer because there is no vacuum.

Ballooning a vacuum panel. Once you have finished a

soundproof panel (created a vacuum within the panel sandwich), you have no way to know that the vacuum is intact unless you place a verification device inside the vacuum area. One simple way to do this is to place a small, high-quality, clear balloon inside the panel before you create the vacuum. Put only enough air inside the balloon to unfold it. As the vacuum outside the balloon is created, the air inside the balloon will expand, inflating the balloon. You can verify the integrity of your vacuum panel by checking the balloon: if the balloon is inflated, the vacuum is intact. If the balloon is deflated, the vacuum is spoiled. (The balloon itself may have spoiled the vacuum if the air inside the balloon has leached through the balloon wall or escaped through a faulty seal into the vacuum.)

Installing permanent vacuum hoses. You may wish to guard against the potential decay of the vacuum within your panels by permanently installing the vacuum hose to the panel and placing the valve and vacuum pump connector on the inside of the room. In this way, you can connect the vacuum pump to the hose, open the valve, and exhaust any air that may have violated the vacuum, whenever you wish. There is plenty of room between the panels and screening devices and the window ledge for the vacuum hose to pass.

YES, WE DID OVERKILL THE WINDOW

The thoroughness with which we dealt with the window is consistent with a *bugproof* room installation. As you will see, the rest of the conference room was modified to bug-*resistant* standards, commensurate with the risk level and the threshold of difficulty we wanted to establish.

We justified the window overkill because one of the three attorneys was afraid of being seen (through the window or anywhere else) with some of the street criminals he represented. He was especially afraid of having pictures taken of himself with any but prominent, headline-quality clients—of the sort that might further his budding political career. The other two attorneys were merely concerned with audio surveillance.

It doesn't matter whether fear of surveillance is based on reality, paranoia, or even some nonsecurity hidden agenda. Since the purpose of the bug-resistant room (including a bugproof window) was to give the three attorneys a feeling of comfort and security and—since it was their money—the modest amount of extra time and expense to upgrade the window was justified.

Word of the bugproof conference room soon spread among persons needing criminal defense, and the law practice grew as a result.

When in doubt, always upgrade the antisurveillance threshold by escalating the level of protection, just as we did with the conference room window. What we did may have been more than was needed, but with personal security measures, more is always a whole lot better than not quite enough to do the job.

CHAPTER FOUR

THE REST OF THE CONFERENCE ROOM

WALLS, DOOR, CEILING, FLOOR, FIXTURES, AND FURNITURE

So far in Robert's suite, we had removed the phone, disconnected the Muzak, and made the window bug-resistant, but not bugproof.

The door, walls, ceiling, floor, fixtures, and furniture had to be dealt with next, but first we had to establish some threshold definitions. A major drug dealer or contract killer would want the best possible security—a bugproof room. Robert and his associates agreed that a bug-resistant room would be good enough for them. The question then became one of threshold (i.e., *how* bug-resistant). At what threshold level of security would the associated attorneys feel comfortable?

The Door

Room access appeared to be a key element. By restricting access to the room to the three attorneys and not allowing other people to be alone in the room, there would be no unaccompanied opportunity to install a surveillance device. They also recognized that a visitor, even though accompanied by one of them, could conceivably place a bug in the room without being observed. As a result, the attorneys decided to account for each visitor who was admitted to the room to help identify anyone who might have placed a device if one were discovered later.

To track visitors, a guest book was placed on the table in the room. The date, time in, time out, and name of each guest and the accompanying attorney would be entered. Code names were to be used for each guest to avoid the risk of the visitor log being stolen or copied.

To prevent unauthorized entry to the room, a five-digit push-button electric lock was placed on the outside of the door. Only the three attorneys knew the five-digit code. To gain access, one of the three attorneys would enter the code on a ten-key pad. The correct code would activate a solenoid that would unlock the door for five seconds during which time the door could be opened from the outside. The solenoid-activated latch was electrically powered from inside the room and could be manually opened from inside the room, but not from outside. In an emergency, the door could be opened by breaking a seal on the key pad and physically removing it from the door.

The wooden door frame was replaced with one made of metal. The door itself was reinforced with metal on the inside and rehung to open outward.

Limiting access to the room also meant that all cleaning and maintenance must be performed by the three attorneys. To allow cleaning or building maintenance people into the room, especially unsupervised, would make the expense and inconvenience of the door meaningless.

The attorneys discussed the installation of a sensor-activated single-frame Super-8 movie camera to photograph each person who entered or left the conference room but decided against it because they did not want to create a film record that could be stolen, copied, or subpoenaed. Such a record would void the code-name protection offered to sensitive clients.

The Ceiling and Floor

Since the conference room was located on the fifteenth floor of a high-rise metropolitan office building, the floors (and, therefore, the ceilings) were constructed of pre-stressed concrete slabs that were bolted to steel I-beams. It seemed unlikely that the ceiling or floor of the conference room would be drilled for a surveillance device from above or below.

There was, however, a thin carpet pad and carpet on the floor and a false ceiling into which light panels had been recessed. While these out-of-sight areas offered locations in which surveillance devices could be hidden, audio or video information gathered by such a device would have had to be recorded in place or carried out of the room over hard wire or by transmitter.

To constantly guard against electronic transmission from within the conference room, I installed an all-frequency radio scanner. The electrically powered scanner was modified in two ways:

1. It was permanently "on" whenever the conference room lights were turned on; and,
2. The speaker was disconnected (so it couldn't be used as a microphone), and a flashing red light installed in its place.

Exterior wall electronic screening (described below) prevented outside radio and television transmissions from activating the scanner.

As a result, even a feeble transmission from any AM or FM signal from inside the conference room would cause the bright red light on the scanner to flash and alert everyone in the room.

To prevent multiplexing over the electrical system in the conference room, all regular electrical power to the room was disconnected. A single 120-volt 60-Hz power line was brought into the room and connected to an electrical "spike suppressor," small transformer, and electrical "filter." All electrically powered devices and lights in the conference room received power from this source.

Electrical spike suppressors are used to protect delicate electronic equipment such as computers or VCRs from power surges. Power filters are used to "clean" the quality of electrical power that operates similar delicate equipment. The small transformer, often built into either the spike suppressor or filter, further controls and insulates delicate equipment from the inconsistent peaks and valleys of commercial and industrial power supplies.

Multiplexing—the use of regular electrical lines to transmit signals—is inhibited by filters and surge suppressors. Multiplexing will not pass through a transformer.

Thus, even though it was possible to hide a surveillance device inside the conference room, it was no longer possible to transmit the information without activating the all-frequency scanner, or to multiplex it over the original hard-wired electrical system.

Walls

Since radio and television signals could enter the conference room through the exterior wall in which the only window (now protected) was located, it was necessary to remove the oak paneling and cover the rest of the bare exterior wall with the same type of grounded copper mesh used as a blocking device in the window.

Once the all-frequency scanner confirmed that no signal radiation was entering the room from the outside, the oak paneling was reinstalled. Since ambient radio and television station broadcast signals are much more powerful than signals broadcast by any surveillance device, we were confident that the conference room was secure from electronic snoops. Unfortunately, that did not protect the room from through-the-wall devices on the other (inside) walls.

We tested the soundproof quality of the room by placing

a Walkman-type "blaster" radio in the room, shutting the door, and trying to hear it with the help of a stethoscope. Once more, with the clandestine help of the night maintenance supervisor, we checked the one conference room wall that was common with a neighboring office. We determined that the base vibrations of the blaster radio could be heard (and felt) but that normal conversation was impossible to detect.

The two inside walls shared by other rooms in the law office were much less soundproof. To further dampen the natural sound resistance of these three walls, we did the following:

1. We removed the oak paneling and glued a ½" thick layer of "bubble" packaging material—the plastic encased material that uses thousands of little air bubbles to cushion and protect delicate equipment during shipment—to the wall.
2. We covered this material with 4' x 8' sheets of ¹⁄₁₆" clear acrylic skin.
3. We glued a second ½" thick layer of bubble packaging material over the layer of acrylic skin.
4. We reinstalled the oak paneling.

We completed the work in less than five hours and tested the sound-dampening quality with our blaster radio and stethoscope again. Although we could feel some of the bass notes, we could no longer tell that the sound we felt was created by loud music.

Fixtures and Furniture

We removed the heavy wood table and chairs and replaced them with a modern clear-glass conference table and matching clear acrylic designer chairs. We also removed the comfortable (but vulnerable) overstuffed leather couch and the two oak end tables and brass lamps.

It is essential to be able to visually check fixtures and furniture for surveillance devices. (It's hard to hide a small tape recorder in a glass table or transparent acrylic chair.)

PERSONAL SURVEILLANCE DEVICES

Although the conference room itself was now adequately secure from audio and video surveillance devices, conversations inside the room were still vulnerable to recording devices hidden in purses, briefcases, or clothing or taped to the skin of visitors.

Unfortunately, the only way to avoid a carry-in device is to forbid purses, briefcases, and other packages; then strip-search each person and replace his or her street clothing (including shoes, bra, panties, watches, earrings, and other items) with presearched "clean" garments—inside the conference room, while you watch—and remove everything else from the room, including your own clothing, to be sure nobody has "planted" a miniature recorder in your own pocket!

Note: Although the all-frequency scanner will warn the occupants if a guest is carrying a transmitter, it will not detect a hidden recorder. It is generally a good idea to keep purses and briefcases outside the bug-resistant room and to furnish any pencils, pads, calculators, or other supplies that may be needed.

Since the clients represented by the law firm wanted to protect their secrets as much as the attorneys, carry-in tape recorders were not considered to be a serious risk.

These estimates and conclusions were appropriate for Robert and his associates because they were attorneys. Nearly all of the "secrets" divulged in the conference room were told to them by their clients. As a result, there was little motive for a client to tape-record his own secret. Robert and his associates were only vulnerable if they, as attorneys, conspired with a client to break the law, offered illegal advice, or acknowledged that funds received by them came from criminal activity.

★   ★   ★

Now that you know how to make a bug-resistant room, let's examine the difference between rooms that are merely bug-resistant and those that are bugproof. We'll see how to build a room that cannot be bugged *while you're using it* or while you're away *without your knowing it*.

CHAPTER FIVE

THE "PERFECT" BUGPROOF ROOM

EARTHING, OR MR. FARADAY'S CAGE

For centuries scientists who work with delicate instruments have been plagued by stray electrical fields. The development of radio and other radiating devices has added greatly to such natural sources of electrical disturbance as sun spots, electrical storms, and so on.

In the 1800s, a scientist named Michael Faraday concluded that he could protect his experiments and his equipment from the influence of these ambient electrical variations by *earthing* the space in which his equipment and experiments were conducted. This earthed, or grounded, space became known as the "Faraday Cage."

Here is the essence of the Faraday "field-free" or "zero-field" theory:

The crust of the Earth is mostly made up of material that is moderately conductive to electricity. In addition, water contained in the ground includes salts and, as a result,

47

forms an electrolyte that conducts electricity even better than the surface. This combination lets electric currents pass from one point to others, through the soil, whenever there are different voltages at different points on the surface of the Earth.

Since this happens at about the speed of light (approximately 186,000 miles per second), it means that any differences in voltage will be instantly equalized. To explain these discoveries and make them useful measurements, names needed to be invented. To do so, scientists assigned the value "zero" to the Earth itself so that all other measured values would be more (+) or less (−) volts, etc.

For our purposes, this means that electrical equipment that is "connected" to the Earth is "grounded" or "earthed" and, as a result, takes on the same value (zero) as the Earth itself. That is, once an object with a potential to carry electrical current is grounded, no difference in potential electrical voltage can exist within it. From a day-to-day practical point of view, grounding an electrical object protects us against electrical shock. That's why modern electrical outlets always have a third post—or "ground." (If you touch a live or charged metal object that is not grounded you will get a shock.) Likewise, radio transmissions—which also are a form of electromagnetic energy—are instantly absorbed into the Earth through the grounded metal screens or plates with which they make contact.

The theory of grounding is key to building a bugproof room. (We used the grounding theory when we constructed an electronic blocking device by installing fine copper mesh screens over the conference room window—and grounding the screens.) Electronic fields on both sides of the grounded screens will be instantly absorbed (converted to zero potential) if they come in contact with the screen mesh or the electromagnetic fields between the wires of the mesh.

Note: Radio transmission inside a carefully grounded room cannot pass through the zero field created by the grounded screen.

While the scientist wants to protect his instruments and experiments from outside electrical influences, we need to keep hostile surveillance transmissions inside the room from getting out. Fortunately for us, the Faraday Cage, which was designed by scientists to keep ambient electrical fields out, works equally well to keep the radio signals transmitted by bugs from escaping.

HOW ELECTRONICALLY "SAFE" ROOMS WORK

If the floor, ceiling, walls, windows, and doors of a room are surrounded by wire netting (or metal plates) and all of them are connected together and then grounded, the room takes on the same potential electrical value as the Earth, zero. Since there can be no potential difference between any one point in the room (zero) and any other point in the room (also zero), no electric field (no radio signal) can be broadcast from inside the room.

The better the grounding, the safer the room from surveillance transmitters. To be perfect, a bugproof room must have its own ground. The best of the available practical grounds is a low-resistance (thick or heavy gauge) copper wire that is directly connected between the Faraday Cage (the metal mesh screens that surround the room) and metallic water supply pipes (which themselves are connected to other water pipes, some of which are buried deep underground). See Fig. 5.

Meetings in a Vacuum

Remember the example of the high-school science teacher who placed a spring-wound alarm clock in a bell jar, then vacuumed the air from the jar? The alarm clock sounded, but the students who witnessed the experiment couldn't hear the alarm because there was no air to carry the sound from the clock to the surface of the jar.

Obviously, we can't live in a vacuum. Our blood pressure would make our body explode, and we couldn't breathe—to mention just two problems. And, of course, we couldn't

FIGURE 5
FARADAY CAGE AND GROUNDS

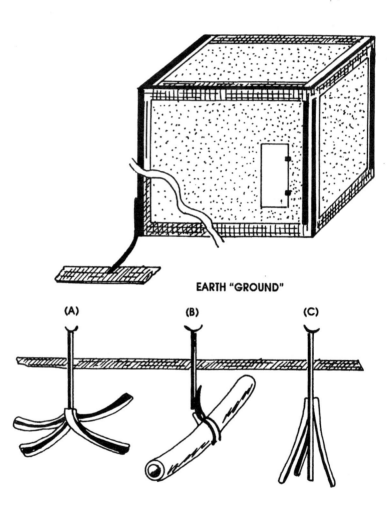

EARTH "GROUND"

A Faraday Cage is any field-free space—whether one-foot square or room size. Protection from surveillance is created by lining the walls, ceiling, floor, and door with copper mesh screens, grounding all screens together, and the entire room to "earth" via (A) earth electrode strips, (B) metal water pipes, or (C) earth electrode rods.

carry on a conversation. Carrying on a secure conversation is, after all, the purpose of a bugproof room, isn't it?

But what if we had two bell jars, one larger than the other. And, what if we were inside the smaller bell jar when the air was vacuumed out of the larger bell jar! We'd be okay, at least until we ran out of air, because the vacuum would only remove the air between the two bell jars.

That is the principle behind the soundproof vacuum panels we installed in my friend's conference room window.

A Room, within a Room, within a Room

Room 1: The outside room is an empty shell, It is the container within which we plan to construct a safe place—a room that can't be bugged while we're using it. Select the place in which you construct your bugproof room carefully. We'll discuss this problem in more detail later.

Room 2: The inside room is the Faraday Cage—the "zero field" room created by covering the walls, floor, and ceiling of the empty shell with fine metal mesh screens or plates and grounding the plates directly to metal water supply pipes.

If you recall the process we used to create the electronic blocking device for the conference room window, you already have almost all the information you need to ground an entire room (create a room-size Faraday Cage).

Caution: While it is true that a Faraday Cage will prevent radio signals from escaping, persons who wish to bug your conversations will use exceptional means to attempt to violate the integrity of your safe cage. This is most often done by probing the cage. That is, a heavily insulated antenna with a noninsulated tip is inserted between the wires of your mesh screen. The tip of the antenna—which is inside the Faraday Cage—will receive the signal and carry it through the insulated antenna shaft outside the cage where it can be recorded or transmitted.

You can protect yourself from this risk in three ways.

First, you will be inside the soundproof room (Room 3, facing page). Second, install the grounded metal screens in a special way. Third, use special electric fans. It is best to use all three safeguards.

First, before you install the grounded metal screens on the walls, ceiling, floor, and door(s) of your room, it is helpful to paint all surfaces with a white rubber-based paint. The white paint will help you see any probe holes or devices that have been placed in the finished room. The paint by itself will not deter surveillance, but it will help later when you illuminate the transparent room with outside lights.

Install all grounded metal screens (except the floor screen) on brackets that stand out at least one inch from the wall or other surface you are covering. In this way, any antenna or other probe that is inserted into the room will be easier to detect since it must pass through the outer wall, the inch of open space between the wall and Faraday screen, and at least one additional inch inside the screen.

Faraday Cage doors: Since you and your associate(s) must enter and leave the Faraday Cage, you must include at least one door. When a door to the room is open, the shielded portion of the room continues to be protected *but radio signals will pass through the open door.* (Even though signals passing through an opening, such as a door, in a Faraday Cage will be distorted, they can be received and unscrambled.)

To be secure, the door itself must be covered with the same screening material as the rest of the room. (The screen on the door should overlap the screen that covers the door frame so there is no unprotected "gap" through which a radio signal might escape.)

The door must be grounded to the rest of the room. You can do this easily by attaching two flexible cables (such as the cables that connect a car battery to ground) one near the top and one near the bottom of the door, on the hinged side. Connect one end of each cable to the grounded screen on the wall and the other end to the screen mounted on the door. Be sure there is enough slack in the cables to allow the door to be opened and closed without stressing the connection at either end. (It is a good idea to solder each cable to the screens at both ends.)

Note: It is essential that the door(s) to the Faraday Cage (zero field room) be closed and the screen covering on the door(s) properly grounded when the room is in use.

Room 3: The third room is the soundproof room we're about to design. When you and your associates are inside this room and the door is sealed, nobody will be able to hear what you say. You already have most of the information you will need to build Room 3.

Soundproofing an Entire Room

Chapter 3 explained how to create soundproof rectangular panels to fit a window. That same procedure works equally well for building soundproof walls.

The main difference between a bugproof room and a bug-resistant room is that in a bugproof room, all walls, ceiling, floor, and door(s) must be sealed against electronic and sound surveillance. Making an entire room soundproof involves making a soundproof door, air-buffeting the external walls, and constructing an acrylic floor that will support the weight of people, chairs, and a table.

Since the inner chamber (Room 3) of our bugproof room will be airtight when the door is shut, we must limit the number of people (air-breathing bodies) who are inside the room and how long any or all of these bodies stay inside the room. My physician gave me the following formula:

Air at sea level is approximately 20 percent oxygen. A normal adult, seated and calm, will consume about 400 cubic centimeters with every breath. Based on normal rates of breathing and oxygen content, it takes about 4 cubic feet of space per person per minute to function. However, if the discussion is tense or any of the per-

sons in the room becomes angry or excited, each person can consume the oxygen from as much as 8 cubic feet of air each minute.

Because the inner chamber is sealed and the air is not replaced while the door is closed, the size of the room will dictate the number of people who can be inside the room and how long they can stay. It is best to allow at least 5 cubic feet of space for each person per minute. That is, allow an empty air space measuring 5' x 1' x 1' for each person for every minute the meeting will last. Keep in mind that bodies and furniture inside the room will displace some of the air. This displaced air must be subtracted from the total cubic foot volume of the room.

Warning: If the oxygen in the air is exhausted, everyone in the closed room will get dizzy and possibly pass out. If they only get dizzy, they will leave the meeting angry and with a headache caused by oxygen starvation. If they pass out, you and they will not be able to open the door and get out of the room, and you and they could die.

Some material may be toxic or oxygen absorbing or contaminating. Depending on the product you buy, who made it, when it was made, how long it has been in storage, or the shared environment in which it was stored (or used), it is possible that the acrylic, bonding glue, silicone gel, and other materials may either deplete or contaminate the oxygen content of a closed room. Likewise, there is less air (and oxygen) on hot days, at high altitudes, in a polluted city, and under other circumstances.

The Importance of Being Canary-Wary

Miners learned long ago that the respiratory systems of canaries and other small birds are more sensitive to harmful gases and lack of oxygen than the human system. That's why canaries were often taken into mine shafts. If the canary got sick or died, the miners ran for safety.

While an outside timing device will be suggested later, such devices can fail or go unnoticed in the heat of debate. Canaries never fail. If you are going to build a bugproof room, you also need to buy or build a transparent nonmetallic bird cage and a (chirping) canary or two. The bird cage must be transparent so it can't be bugged.

Put your canary in your cage and take it with you into the inner chamber. If the canary gets sick—break the seal on the door and get out of the room! A sick canary means somebody has gassed your room or you've run dangerously low on oxygen.

Always test the closed room with a canary before you expose people. It may not be nice to make a canary sick or kill it, but if your room goes bad, your business associates and clients will never forgive you—if they survive! And if you aren't in the room with them and they survive, they'll think you tried to kill them. That is why there is only one latch and it is on the inside. No one wants to be locked inside a transparent airtight room—especially when he can watch the person who locked him in smile as he slowly suffocates. (Lock the Faraday Cage from the inside and don't leave objects that can be used to block the door lying around.)

How Big a Room?

Since the purpose of the room is to hold secret conversations, it is important to put the horse before the cart and first determine how many people will meet and how long they will be inside the room in order to determine the appropriate room size.

The chart below may be useful as an information guide to help you estimate the size of a room you may need.

Space for a 5-10 Minute Meeting

Persons	Minimum Air Needed	Minimum Room Size
2 persons	100 cu. ft.	6 x 6 x 5 feet
3 persons	175 cu. ft.	6 x 6 x 8 feet
4 persons	350 cu. ft.	6 x 8 x 8 feet
5 persons	475 cu. ft.	6 x 8 x 10 feet
6 persons	650 cu. ft.	6 x 10 x 12 feet

This chart is furnished as an information guide only. You must carefully calculate the size of any room you design or build based on the specific conditions at your location. If you are not scientifically qualified to estimate necessary air quality, consumption data, and available volumes, consult someone who is both professionally qualified and familiar with local conditions. Local conditions include, but are not limited to, altitude, temperature, and air quality. When in doubt make the room larger, the number of people fewer, and the time inside shorter.

Structural considerations make it difficult and expensive to construct inner soundproof chambers larger than 10 feet wide by 12' long by 8' high. In practice, the inside height tends to be 6'6".

Plan meetings to last ten minutes or less. If a meeting lasts longer than ten minutes, it *must* be interrupted by five-minute breaks every seven to ten minutes. During breaks, everyone must leave the room so the stale air inside the room can be exhausted and replaced by fresh air. If more persons are inside the room than the size allows, the time must be reduced.

Making It "Perfectly" Clear

Every component of the soundproof inner chamber—Room 3—must be constructed of clear acrylic sheets. Every piece of furniture must be constructed by you of clear acrylic sheets or assembled by you from clear glass, plastic or acrylic material: glass table legs and plastic or acrylic chairs and table top. If you purchase a finished chair or table, select one that is completely transparent. If it isn't, take it apart and replace as much of the opaque (nontransparent) material as possible. If any object you buy uses metal screws or parts, replace as many of them as possible with clear acrylic screws or bond the components together with transparent glue. *Remember:* it is difficult to hide a transmitter or recorder in a transparent table or chair.

Thus, even though someone may violate your grounded Faraday screen with an insulated antenna probe, it won't do them any good if there is no place to hide a transmitter inside the soundproof inner chamber.

Build the Floor First, Everything Rests on It

Years ago the airline industry had an expensive problem with the aisles of their new jetliners. Passengers kept punching holes in the aisles! It took millions of dollars to replace the original aisles and resolve the problem.

The original aircraft aisles and seating area were built of lightweight "honeycombed" metal—much like the walls of corrugated cardboard boxes. Thousands of these short vertical corrugated metal walls were bonded together to make a lightweight honeycombed deck. The deck was more than strong enough to support the heaviest passenger or food cart. But the problem didn't come from 500-pound Japanese wrestlers or food and booze carts. It came from 102-pound women in high-heeled shoes.

The weight per square inch of a 102-pound woman when she walks on a shoe with a ¼″ spike heel can be 1,632 pounds per square inch. So, even though the honeycombed decks of the jetliner could handle a 1,000-pound food cart, they broke under the pressure of petite women in spike-heeled shoes.

The soundproof floor of the inner chamber won't be built of honeycomb metal, of course, but it is vulnerable to cracks. Even though the acrylic skin probably won't break or shatter, it might crack under stress and release the vacuum. No vacuum—no bugproof room.

As a result, you're going to have to spend some extra money on the floor portion of the bugproof room. Even though you construct the vacuum floor panels much like the ones we made for the conference room window, you will have to use smaller triangular sandwiches, use more of them, place them closer together, and cover them with two layers of ¼″ acrylic skin when you assemble the floor.

This will take more time and a little more material, but it won't cost much more since the materials are the same (¹⁄₁₆″ and ⅛″ acrylic) as we used to make the conference room window. The ceiling and walls are constructed entirely from this ¹⁄₁₆″ and ⅛″ material.

Only the two skins of the floor, the footings, and the door will require ¼″ thick acrylic. One-fourth-inch thick 4′ x 8′ sheets of clear acrylic cost as much as $100.

You will need to bond two of these ¼″ thick sheets together to make each outside (½″ thick) skin for the floor. Except for the added thickness of these outside skins, the floor panels are the same as the panels for walls and ceiling.

You will also need floor footings to support the bottom of the deck about 2 inches above the mesh screen that covers the painted floor of the room. (The mesh screen on the floor does *not* need to be separated from the floor by brackets as it is on the ceiling and walls.)

Each footing can be prefabricated by bonding 4″ x 4″ wide squares of ¼″ acrylic to make a hollow square 2 inches high. The assembled hollow square should then be capped at both open ends by a 12″ x 12″ x ¼″ acrylic square. (Glue the 2″ high 4″ x 4″ square to the center of the 12″ x 12″ square. See Fig. 6.) Each square is glued to the metal mesh on the floor *but not* to the acrylic floor that will be placed on top of the footing.

The shifting weight caused by the movement of persons in the room will stress the deck and cause the floor to creep slightly. It is important that the bottom surface of the acrylic floor be able to adjust without transferring the stress to the

FIGURE 6
FOOTING SUPPORTS

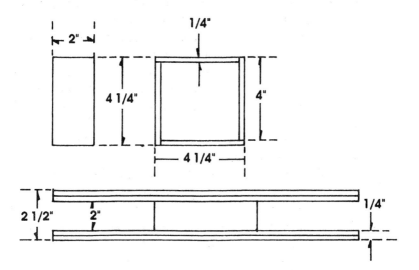

footing and possibly breaking the footing or cracking the floor skin which, in turn, might void the vacuum inside the floor panel.

Six to eight footings should be adequate for most bugproof rooms. Extra footings will make a more solid deck, generate a more secure image, and reduce the chance that the floor will warp or crack.

You may wish to cover the inside floor skin on which you walk with $\frac{1}{16}''$ thick clear acrylic sheets. If you do, don't bond the sheets to the $\frac{1}{2}''$ thick acrylic inside panel skin you constructed. The $\frac{1}{16}''$ sheets will act as a carpet to protect the much more expensive stress-bearing inside panel skin from abrasions and maintain the transparency required to check for newly planted bugs or surveillance gadgets.

An Airtight Hatch

The door and door frame are easy to make. The actual door doesn't have to be large if you place the transparent chairs and table inside the room before the final wall is installed. You probably won't have enough clearance to bring furniture into the room through the ceiling, which can't be installed until the first three walls are in place.

The overall height of ceilings in most rooms is eight feet. You will lose the use of about six inches of space next to the floor of the room because of the wire mesh, footings, and the deck of the inner chamber. You will also lose about six inches from the ceiling to brackets, the Faraday mesh screen that hangs from them, and the soundproof ceiling panel of the inner chamber. As a result, you will have about seven feet of vertical space available.

You should plan to make the inside height of your inner chamber about 6'6". This will allow comfortable head room and give you a safety margin of air volume. Since the height of the wall panels will be uniform, you can have the clear acrylic sheets (for the walls) precut to 6'6".

Assemble the soundproof wall panels exactly as we did for the conference room window panels. Since the soundproof room will be longer and wider than four feet, you must overlap the ⅛-inch outside skins, as the panels are built to assure a proper vacuum seal.

Likewise, when two or more panels are placed side by side, the abutting sides must be included in the vacuum. Not only must the common panel edges be glued together where they join and the joints caulked with silicone gel on both outer surfaces of the panels, an additional ¹⁄₁₆" x 12" x 6'6" strip must be glued over both sides of the seam between the panels. It is prudent to caulk the edges of these carefully glued seam overlay skins with silicone gel.

Note: You must decide whether or not to make each wall, floor, and ceiling panel a self-contained vacuum. If each panel is to be separate, each must have its own valve. Not only is that more work and expense, it means there can be no vacuum between the panels. It is more effective to make each wall, ceiling, or floor a contiguous unit with a single vacuum valve. Thus, before joining panels to form a common wall, you must drill matching airway passages between panels that abut and be sure the skin-strips that cover the seams between panel sections are airtight. (Please review the procedure for installing valves and creating panel vacuums in Chapter 3.)

The Port of Entry

The door should be placed close to the corner of a side and end wall, with the door frame built into the side wall and the door hinged from the outside onto the wall at least six inches from the end.

Although you can cut the door opening from a finished side panel, it is easier to allow for the door when one of the side panels is constructed. The door should be at least 2 ½' wide and 4 ½' high. The bottom of the door opening should be at least 6 inches above the bottom of the panel, and the top of the opening should be at least 6 inches below the top of the panel. (Picture the door "hatches" that separate compartments on a navy ship.) To enter or leave the inner chamber everyone will have to step over a 6" threshold and duck their heads to pass through the door.

Square corners on the door frame and the door work just as well as fancy curved doors and are a lot easier to build. You already know how to build the panel. The only difference in the panel that contains the door frame is that the three unhinged surfaces must be at a slight angle so the door will seat properly. A matching angle is also needed on the three unhinged sides of the door itself. (See Fig. 6.)

Whatever the size of the door frame you design, the edge inside the room should be about three inches smaller than the exterior dimensions. Likewise, the complementing dimensions of the door should be about 3 ½ inches smaller than the door frame into which it will seat when closed. To be complementing, the smaller dimension on the door should be on the inside of the room.

In addition, a ¼" thick x 2" wide acrylic strip should be bonded to all four sides of the door frame to form a lip on the *inside of the room*. The purpose of this lip is to overlap the door opening by at least 1 inch.

A transparent polyurethane gasket (similar to the refrigerator door seals used in the conference room window) should be mounted on the door side of this lip so that the door, when closed, will compress the polyurethane gasket and create a seal. Additional polyurethane seals may be used elsewhere, as needed, to assure the tightest possible interface between the door and the door frame. Keep in mind the need to inspect everything by seeing through objects, including polyurethane gaskets.

As a further precaution, you may wish to use petroleum jelly on the polyurethane to make the same kind of airtight seal that was demonstrated in the science teacher bell jar illustration. Even though the polyurethane seal with a thick coat of petroleum jelly does not create a vacuum between the door and frame, it will make the interface airtight and sound absorbent. That is a very effective way to make the room soundproof when the door is closed.

Hanging the door and latching it when closed is best accomplished if you can find transparent hinges. *Use no screws* to attach the hinges or latching device directly to walls or doors. The only surface available into which you can screw a hinge is a vacuum panel. As a result, the screw will jeopardize the integrity of the vacuum and may void the soundproof wall or soundproof door panel.

Glue works wonders. If you can't get the glue to hold the weight of the door, try to use clear acrylic screws, and then only screw the hinges onto a separate ½" thick strip of acrylic, which you will bond to the outside of the panel and door with glue. If you cannot find transparent acrylic hinges or screws, use aluminum products. Since they are mounted on the outside of the soundproof wall and door, they will not transfer conversation from inside the room. (See Fig. 7.)

Closing the Door Behind You

Latching the door shut is like closing the door of a horse barn. You may be able to find a ready-made transparent

FIGURE 7
INNER-CHAMBER DOOR AND FRAME

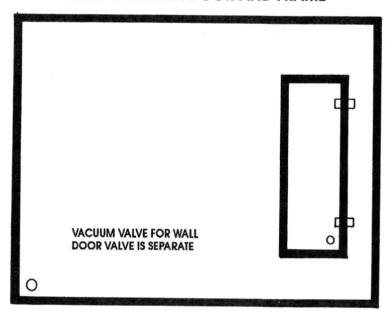

**VACUUM VALVE FOR WALL
DOOR VALVE IS SEPARATE**

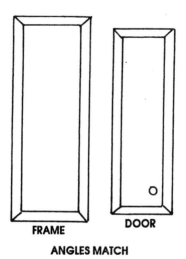

FRAME **DOOR**

ANGLES MATCH

latching device. If not, you can design one similar to the following:

Make two U-shaped latch keepers:

Part A: Bond two ¼" x 2" x 3" strips together to make a ½" x 2" x 3" strip. Make six of these strips.

Part B: Bond two ¼" x 2" x 1" strips together to make a ½" x 2" x 1" strip. Make two of these strips.

Glue two Part A units to one Part B unit to form each square U-shaped latch keeper.

Glue one latch keeper near each side of the inside of the door, at the same height, so that the end of each latch keeper is at least 2 inches away from the edge of the door and does not interfere with the movement of the door when it is opened or closed. Mount the latch keepers so the closed end (bottom of the "U") is down.

Make a latch bar: Measure the door frame (opening) and make the length of the latch bar at least 4 inches longer. The end of the latch bar on the hinge (end wall) side of the door may be less than 4 inches but must be at least one inch beyond the door frame.

Cut two ¼" x 2" strips of clear acrylic to the length you just measured. Glue the two strips together to form the basic ½" thick latching bar. Now glue a 2" strip of polyurethane gasket across each end of the latching bar. The polyurethane gasket will prevent the latching bar from scratching the door frame and will keep the door tightly shut.

Door closing fixture: To close the door from the inside, you will need a handle or other fixture to pull the door shut. Any transparent fixture will work. *Do not use screws* to mount it to the inside of the door. If you can't find a glass or clear plastic or acrylic door handle at your hardware store, you can make one by stacking four different sized ¼" thick rectangular acrylic pieces together and bonding them so the smallest piece is glued to the inside of the door and the largest piece is farthest from the door. That is, glue a 2" x 2" x ¼" piece to the door, a 3" x 3" x ¼" piece to the first piece, a 4″ x 4″ x ¼″ piece to the second piece, and so on.

To close the door from the inside, pull on the door handle until the door is properly closed, then insert the latching bar into the latch keepers mounted on the inside of the door. When you insert the latching bar into the latch keepers be sure the polyurethane gaskets are away from you and toward the door.

FIGURE 8
DOOR LATCH KEEPERS AND BAR

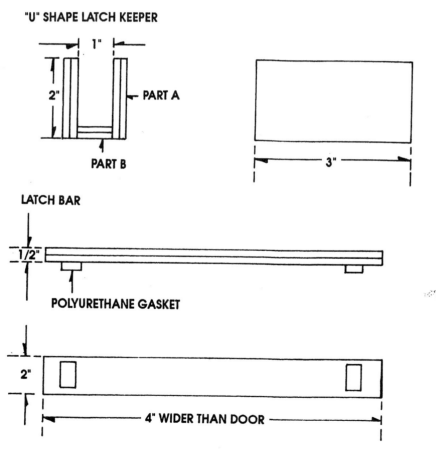

"U" SHAPE LATCH KEEPER

1″ — PART A — 2″ — PART B — 3″

LATCH BAR

1/2″

POLYURETHANE GASKET

2″ — 4″ WIDER THAN DOOR

MAINTAINING A STATE OF EMPTINESS

The door, each of the four walls, the ceiling, and the floor of the soundproof room all contain a vacuum. Each must have its own valve, and each must have its vacuum formed and sealed separately.

Even so, there is some porosity in nearly everything, and, over time, air will find its way into the vacuum through the silicone gel and glue and possibly through flaws in the acrylic skin itself.

It may be desirable to spend $600 or so to buy a professional vacuum pump because you have a large volume of air to exhaust and you must create as high a vacuum as possible. While duct tape on a vacuum cleaner may create a satisfactory vacuum in a 3' x 4' double window panel, it won't pull enough vacuum to make a 6' x 10' x 12' single wall completely soundproof.

Although a way to test for vacuum integrity (without using balloons) will be presented next, it is a good idea to upgrade the vacuum within the door, walls, ceiling, and floor once a month whether you observe signs (sounds) of vacuum decay or not. Owning or renting a professional vacuum pump will be money well spent. (See Fig. 8.)

Lights, Fans, and Other Gadgets

You will need three fans. One should be at least 18 inches in diameter and mounted on a stand. Each fan must have three or more blades. (I bought a good one with three speeds for less than $25.) Even though you will use the first fan to recycle the air inside the soundproof chamber before and after each five to ten minute use, it also serves to cool the room and will be used to create a random noise. This artificial wind-noise has several uses. Among others, it helps cover any accidental comments made when the door to the soundproof chamber is open.

To make the noise unpredictably random, use a carpenter's file to notch the front of one of the blades three times. Notch the back of the blade next to it two times. Do not notch any other blades.

Place this fan facing the door so that it points through the door at an angle and pushes air through the door toward the center of the room. This first fan should be about three feet from the door. *Turn this fan on high and leave it on the entire time you are inside the Faraday Cage*, whether you are inside the soundproof inner chamber or not or whether the door to the inner chamber is open or closed.

The second and third fans should be different makes or models and should have a different number of blades. Each should have a fan diameter of at least twenty-four inches. Notch the blades of the second fan, but use more or fewer notches on the blades you intentionally damage and always leave at least one blade without any notches. Place the second fan close to the mesh-covered floor at the far end of the wall in which the door is located and aim it slightly upward and at the center of the wall. Turn this fan on high and leave it on during the entire time you are inside the Faraday Cage, whether you are inside the inner chamber or the door to it is open or closed.

Repeat the notching process with the third fan. Place it at the corner of the soundproof chamber diagonally across from the door. Aim the fan so that most of the air is directed toward the wall opposite the door. Some of the air should catch the end wall of the inner chamber. Turn this fan on high and leave it on during the entire time you are inside the Faraday Cage room, whether or not you are inside the inner chamber or the door to it is closed or open.

Because these fans have a different number of blades, operate at different speeds, and have notched blades, they make a lot of noise. In addition, the air they push against the outside walls of the soundproof room will strike the walls in constantly changing, random nonpatterns.

Even though you have made it virtually impossible to hear conversation through the vacuum of the wall panels, the constantly changing wind from the fans will shake the outer walls and greatly overpower any fluctuation that might escape because of a faulty vacuum. The random noise from the fans prevents even the most sophisticated computer from unscrambling any voice-generated sound from inside the chamber. They also serve to constantly test the soundproof status of the inner chamber when the door is closed.

Caution: If you can hear the fans from inside the soundproof chamber with the door shut, the vacuum in one or more panels or the joints between them has failed.

To fix a sound leak, determine where the sound you hear is loudest and check seals. Sound leaks can usually be resolved by bonding a ¹⁄₁₆-inch skin overlay to the offending area and vacuuming it again.

Similarly, you can test for electronic integrity with any sensitive AM/FM radio. If you can hear a radio station, the Faraday Cage isn't working. If you don't have a directional antenna to locate the failed area, place a 2 + watt walkie-talkie inside the Faraday Cage and try to pick up its signal outside the cage. You will hear the walkie-talkie more clearly in one or more areas outside the cage. Patching the wire screen or improving the grounding (earthing) system will usually fix any electronic leaks. It is also possible to accidentally create a zone of vulnerability if the screening material is not identical throughout the cage. If this happens, replace the "odd" screens and reconnect the ground between the new and old Faraday screens.

Lights and Other Gadgets

There must be no electric device inside the soundproof acrylic chamber. That means no lights and no electric outlets. If your meeting involves bookkeeping, *you provide* a battery or mechanically operated adding machine. *Never* let a guest bring his own pocket calculator inside the room.

Since the inner chamber is constructed entirely from transparent acrylic material, lights inside the Faraday Cage room but outside the acrylic chamber will adequately illuminate the inner meeting room. It is best to hang 150-watt soft white electric lights near the ceiling and the four corners of the room. Illumination from above is more nearly normal, and placing a light at each corner of the room will remove harsh and unflattering shadows. If the illumination is too bright, use 60- or 100-watt bulbs. If it is too dim, use more lights. Do not use lights brighter than 200 watts. Be sure all lights are far enough (usually eighteen inches or more) from any acrylic surface so the heat does not affect acrylic seams or soften or distort the acrylic, glue, or silicone gel.

SAFETY

Don't kill yourself over a conversation. Other than the canary, the only absolutely required gadget for your bugproof room is a timer, similar to the type used to make enlargements in a photographic darkroom.

Nearly all of these timers have built-in electrical outlets, which the timer turns on or off. Most good timers cost between $20 and $75, and some of them have normally open circuits. If you can't find one with such an outlet, modify the one you use so that it operates in a normally open condition. That is, instead of keeping a light on during the timed period, you want the timer to keep a light off when it is timing and turn a light on when the time is up. (Be sure the timer can be set for five to ten minutes.)

You can modify a normally closed darkroom timer to operate normally open by adding a solenoid switch. This will let you plug the light into the solenoid switch instead of the timer itself. Then, when you set the timer for five to ten minutes and turn it on, the current from the timer will keep the solenoid open (and the light off) until the time is up. When the timer shuts off, power from the timer to the solenoid will stop, and the light will go on.

Warning: The purpose of the external timer is to warn you that the air in the chamber is no longer safe to breathe, hopefully long before your canary feels the effects.

First, you must be sure the light bulb works each time you set the timer. Second, having two operating bulbs is a good idea. Third, even though you buy the world's best timer, it is outside the sealed room and can't warn you about contaminated inside air. Fourth, do not take a professional air sampling "sniffing device" inside the room with you. They can be tampered with, they may fail, and they are a great place to hide a bug because they are not transparent and you can't take them apart for inspections.

To be sure you notice the warning lights, you should modify the light sockets to make the lights flash. Most hardware stores that sell lighting supplies also sell flasher units for regular 110-volt light bulbs. These look like coin slugs or fuse-box slugs and can be placed inside the light socket between the screw-base of the bulb and the electrical contacts. They function to interrupt the current that passes between the socket and the bulb when the power is on and cause an ordinary 110-volt light bulb to flash on and off to capture your attention. I like to use two 100-watt light bulbs, in side-by-side reflectors, aimed eye-level at the inner chamber. One 100-watt bulb is red, and the other is blue. When the timer stops and the two warning lights start to flash, they seldom flash simultaneously. This "police car" flashing effect *will* get your attention.

Don't do what one fellow suggested: use a loud police siren instead of the flashing red and blue lights. If you've done a good soundproofing job, you'll never hear the siren. Everybody I know pays attention to flashing red and blue lights—even those who ignore sirens.

Finally, one more time: *don't forget the canary!* Once you're inside the soundproof chamber, nobody can open the door from the outside. The door opens outward, and there is no outside handle to pull on. The inwardly angled nesting of the closed door prevents it from being crashed in.

No one will be able to break down the door in time to save you should you pass out! No matter how hard they try.

FAIL-SAFE BACKUP

If that bothers you, build a small, secret, knock-out window into one of the panels. Place it in an obscure place, preferably on a back wall, close to the floor where you can kick it out. Don't tell *anybody* about it.

At least then, if a "trusted" associate decides to block the door and watch through the transparent wall as you slowly suffocate while red and blue lights flash in your face, you can kick the secret window out and breathe.

If you're afraid somebody will shoot you while you're inside the room, build the inside skins of bulletproof acrylic. It's transparent and the best possible material for your walls, ceiling, floor, and door, but it's really expensive. There are some folks who need bulletproof/bugproof rooms—and *can* afford them!

To be of any value, a "kick-out" window must be secret. Read this page, carefully cut it out, and destroy it! If you don't, others may search for the window. If they do, a window will be useless to you should you need to use it! There is no other reference to a "kick-out" window. If you plan to build a "knock-out" window, destroy this page.

CHAPTER SIX

MOBILIZE YOUR BUGPROOF ROOM

FINDING A MOBILE HOME
FOR YOUR BUGPROOF ROOM

If you think about it, there is little about a bugproof room (at least a good bug-resistant room) that can't be built into a self-contained motor home or large customized van.

In many ways, remodeling a motor home is easier. It certainly solves a lot of problems regarding where to put it, how to guard it, and how to maintain a nearly invisible identity profile for all who enter and leave it.

Sometimes, of course, you *must* meet clients or associates in a conference room environment. For most purposes, a fixed location is the only one that makes sense. In addition, motor homes have an image problem.

Many customized motor homes and vans are every bit as impressive and classic as a posh office. And they can be even more expensive than they are impressive. In the final analysis, whether or not an antisurveillance vehicle will work for you is a function of who you are and/or the image you hope to project.

Even if you've got a $70,000 supercustomized, self-contained vehicle with posh appointments, a mobile until won't be as professional in image as an office in a high-visibility building with a prestigious address.

A mobile security room also may have implications you don't want or can't afford. It may imply that you frequently deal with unsavory folks who don't dare show their faces in public, because they are wanted by the FBI or otherwise have a rank aroma about them. If so, you *will* take on the aroma of guilt, by that implied association. Be warned, if the people from whom you want to keep your secrets find out about your mobile security vehicle, you may draw a lot of intensive and immediate special attention.

No Problem for Most Plain Folks

On the other hand, a customized antisurveillance van or motor home can be pleasant, comfortable, convenient, and secure. If meeting in a nice-looking, comfortable motor home with out-of-sight social amenities isn't a problem, modifying one to be bug-*resistant* can be quick and easy. Not cheap—just easy.

Tax deduction? If you have or want a posh motor home or customized van, you may be able to deduct it as a business expense, because it is a necessary mobile security room. If you claim it, do so only if you can afford to have the government know that you are in a business that must keep secrets.

Furthermore, if you are careful with your modifications, nobody outside the vehicle should know that it is special and neither should your visitors—unless *you* want them to know.

Whether or not you tell your clients or associates about the antisurveillance features of your vehicle is a business and security decision that you must make logically. Never boast about it—or any other safety or security system, for that matter. Bragging and other forms of exhibition com-

promise any system and make you look careless and unpro-
fessional.

"Don't Bug Me," Said the Van

Since protecting the vehicle from electronic bugging is
the most difficult problem to solve, you might want to consider not protecting the vehicle itself from electronic
snoops—but parking it in a Faraday garage. Or, you might
want to do both. For example, carry on your secret transactions in the Faraday vehicle anywhere; your top secret
transactions inside the vehicle, only when the vehicle is
inside a Faraday garage.

Otherwise, you can adequately bug-resist a vehicle if
you isolate the conference area from the driving compart-
ment, contain the conference area in a Faraday environment, ground it when you are moving with the kind of
antispark "grounding drags" sometimes used by gasoline
tankers, and with one or more static grounding devices
when you are parked. (See Fig. 8.)

You already know everything you need to know about
vacuum panels and Faraday Cages. Except perhaps if you
want to use an 18-wheeler. I personally don't believe a
mobile unit can be made truly bugproof, but some others
do, and they may be right. Nearly every antisurveillance
expert will agree that a bug-resistant mobile unit is at least
as good technically as a fixed-location bug-resistant room.
To repeat: if a bug-resistant mobile unit is parked inside a
Faraday garage when you hold meetings, the combination
can be bugproof.

No, not an Acrylic Van

The primary reason for building the inner chamber of
a surveillance-free room of acrylic is to make it transparent.
You can sweep a room today, but how can you be sure some
night visitor didn't hide a bug if you can't make a visual
inspection?

That's the big problem with any fixed location. Obviously, you can't take a fixed-location, bugproof room home

with you. Not so with a vehicle, whether it's a large customized van or a full-size luxury motor home. If you can take it home, you can take it anywhere. And anywhere can be a secure place—with plenty of isolation, locks, dogs, and surveillance cameras.

It may help to think of your bug-resistant vehicle as "pure" right after you finish the modifications. You know it remains pure only until you leave it alone with somebody else. ("Trusted" associates are the most dangerous because some degree of resentment lurks in the hearts of most people.) From the day your clean van is finished—and the first minute you leave it—you may not be able to tell if it has been "admired" from a distance, but there are a few traps you can set that will let you know if it's been violated. If somebody *has* got next to it you need to *know it*.

Backup the Backup

You need at least two motion sensors, which don't cost much. Place the primary sensor in a logical place and connect it to whatever you want—a telephone dialer, alarm bell, or lights. Don't connect it to a transmitter unless you use a receiver to activate something obvious: a remotely controlled floodlight, bell, or siren.

Think cause and effect. If a light goes on, there must be a sensing device. Expect the first sensor to be found and neutralized. Typically, an intruder will trip your alarm and flee. He'll watch for a response. When none comes, he'll return to finish the job. You want him to find the first sensor, so he won't look for your backup systems. If he flees without damaging your vehicle and doesn't return, fine. If he does return, you want him to quit looking after he finds the throwaway sensor. Never connect any backup sensor to an obvious device (light, bell, or siren) that is triggered by your throwaway sensor.

Make your first sensor part of a working system and be sure the professional appearance and quality of that sensor is consistent with the professional quality of the property it protects. Nobody out to plant a bug in your vehicle will believe you haven't protected it in some way.

Now put the additional sensors in less logical places. Let them do only one task—transmit to one or more well hidden remote cameras located where they will cover the entire area. Use equipment that will record clear pictures in whatever ambient light will be available should somebody invade the space near your vehicle. This may require twenty-four-hour tamperproof lights, infrared lights, or very low light (VLL) or "Starlight"-capable camera lenses.

Put at least one VCR camera or motion picture camera in a Faraday Cage. A real professional out to bug your bugproof room—fixed or mobile—will degauss the area before he leaves and that will erase all magnetic media (including your VCR tapes) in the area.

I prefer low-light video cameras for instant tape replay because film has to be developed. The camera should have battery back-up and an FM on switch that has a thirty- to sixty-second timed off switch. In this situation, when the motion sensor transmits its signal, the camera is turned on and operates silently for a minute or less until the timer turns it off. If the motion continues, the sensor turns the camera on again. A standard VCR cassette has 360 minutes' worth of protection and, operated in thirty- to sixty-second bursts, should last for at least a month. A Super-8 motion picture camera has the same coverage on film at one frame per second (time-lapse) speed.

The theory, of course, is that the intruder will look for and find the first sensor and disarm it (or flee), while the additional sensors continue to function. Meanwhile the sensors have turned on the camera(s) and you will know who cared enough to visit your vehicle *and what they did.*

If you know you've been bugged, you have lots of useful information. First, you can expect future conversations to be overheard. Second, you can infer that prior conversations have not been bugged—unless your visitor only changed the batteries in an extant bug. Third, you know there is a bug, so you can remove it. Fourth, you know there is a bug, so you can leave it in place and exploit it for self-serving disinformation purposes, as my friend did with his reverse-sting operation. Most importantly, if you've been bugged, you know somebody out there cares enough to put you under surveillance—and that means your "fears" of being watched are real and not a paranoid reaction.

Never hard-wire a sensor to a camera. If an intruder finds the sensor, he will look for and find the camera.

A WARNING ABOUT GUARDS

Never trust a guard to protect or be anywhere near your antisurveillance vehicle or inside a fixed-location bugproof room.

Generally speaking, a guard is bored, underpaid, and too often thinks of himself as some kind of undiscovered private television cop who not only will snoop out of curiosity, but will want to improve his personal image and "professional status" with his buddies by bragging.

I believe most guards are honest and conscientious. That is especially true of men who supplement their military retirement with part-time work. Unfortunately, that's not true of all guards. A few are drunks, some are incompetent, and occasionally one is just plain dishonest—a private cop who looks for targets of opportunity amongst the property he guards. Your property is, after all, valuable enough to need guarding. Don't forget that many guards are informants and not always police informants. All human guards can be bribed, threatened, or intimidated.

When a guard supervisor tells his men to report anything unusual, it too often really means "tell me something I can sell or trade." The supervisor may or may not use the information to his personal advantage. If he does, it is always at your expense. Your antisurveillance vehicle is unusual. Its existence is valuable knowledge that is worth selling or trading. It is clearly an unusual target of opportunity.

Finally, the most a guard can do for you is apprehend somebody or call the cops. In either case, police will be involved.

Do you really want to have police involved? Do you want to get a call from them? Or have them inspect your vehicle? Or have your vehicle become part of official police records? All countersurveillance information is too juicy to ignore. How juicy? Let a cop inspect your vehicle, and the chief will know tomorrow and the mayor the day after—providing they speak to each other.

All law enforcement administrators are bureaucrats, and no bureaucrat will put your best interests ahead of his own!

CHAPTER SEVEN

NOTES, REMINDERS, AND OTHER TIPS

FIRST, TWO LEXICOGRAPHY NOTES

Michael Faraday is the scientist who discovered the "field-free" room concept in the 1800s. Thus we can say that an area inside a Faraday Cage has been "Faradized" (verb); or if we're going to make it safe we are going to "Faraday" it (infinitive form of verb); or if the job is well done one might say, "I like your Faraday" (noun). This room has been Faradayed (verb). Take me to the Faraday room (adjective).

As an author, my ego demands that I name a bugproof room that has been made soundproof (via vacuum acrylic walls, door, ceiling, and floor) and is nested inside a Faraday Cage after me. I'll call it a Glas room (adjective). (Glas with one "s" please.) That is, the room has been Glased (verb); the Glas looks nice (noun).

HOW MUCH IS THIS GOING TO COST?

Experience teaches that the two most difficult problems in actually building a bugproof room involve the acrylic and the vacuum pump.

To check on 1990s prices, I visited a professional glass door and window retailer. This was a large, well-stocked supplier to remodelers and do-it-yourself folks. They maintained an inventory of clear acrylic (optical quality) in $\frac{1}{16}''$ and $\frac{1}{8}''$ thickness only. Their sheets measured 4' x 7'. They did offer to custom-cut the acrylic at no additional charge, but their per sheet price was ridiculous. This glass specialty outlet wanted $144 plus tax for each 4' x 7' sheet of optical quality clear $\frac{1}{8}''$ acrylic. They said they could special order $\frac{1}{4}''$ sheets of 4' x 7' acrylic but warned me that it would be too expensive.

When I complained about their prices, they tried to sell me insulated glass panels and sound-reflecting and sound-dampening windshield-type glass. The truth is that insulated windows or glass similar in composition to automobile windows do reflect and absorb sound. That is helpful, but it just isn't good enough.

While the friendly sound engineer in a recording studio can use his professional equipment to filter out unwanted noise and conversation, the unfriendly surveillance professional will use similar equipment to isolate and amplify the same conversation. The message here is that almost good enough and sound suppression just aren't worth a thing.

The acrylic you need is the most expensive part of building a bugproof room. Never buy from specialty outlets like the glass window and door shop I just described. They not only charge too much, they will try to sell you what they have available, and it just won't do the job you need.

I checked the yellow pages and discovered a nearby building supply outlet. They sold acrylic panels and maintained a large off-the-shelf inventory at the following single-sheet retail prices:

- $20.98 per sheet for 3' x 6' ⅛" optically clear acrylic.
- $9.89 per sheet for 2' x 4' ⅛" optically clear acrylic.
- $2.38 per sheet for 2' x 4' ⅛" clear (not optically perfect) panels, normally used by builders for transparent ceiling light panels.

You should have no trouble finding similar prices and off-the-shelf inventory at a building supply center near you.

It doesn't take much imagination to see that you can bond together inexpensive sheets of ⅛" panels to make everything you need, including the floor and door of a bug-proof room. Even so, you may wish to spend the extra money to special order ¼" sheets (or even ½") material. If so, find a source that will sell you ¼" sheets (4' x 7' or 4' x 8' clear acrylic) for $100 or less.

If you have trouble finding a source for extra thick transparent acrylic, visit any liquor store or check-cashing agency in a high-crime area. Just ask the manager where he buys his bulletproof glass.

In practice, the smaller size (much cheaper) sheets are nearly as easy to use as the 4' x 7' panels because you are going to honeycomb them into acrylic triangles and panels and bond them together anyway.

CONSTRUCTION TIPS

Cutting the acrylic seems to be a big problem for some people. Remember that the smaller the teeth in a saw blade, the finer the cut. Some optically clear acrylic is brittle. Buy a single piece and experiment with it before you invest a lot of money. If you aren't successful with the test panel and can't find (or trust) someone else to cut it for you, use the thinner and/or more flexible sheets. They won't be quite as clear optically, but they will be transparent and you will be able to inspect for hidden bugs and other devices once the panel (or triangle) is in place.

The same is true for drilling the acrylic. If you have trouble, take your sample to a building supply outlet and explain the problem. They will be able to help. (If someone wants to know what I'm making, I usually say I'm building or remodeling a soundproof recording studio.) If you continue to have trouble when you drill holes, downgrade from the brittle, optically clear acrylic to the thinner and more flexible material.

Mounting vacuum valves seems more complicated than it really is. First, find the valve you intend to use. The valve size will determine the size hole you need to drill. Plan to mount the valve on the outside of the hole because the vacuum will be inside and will try to suck the valve into the hole. Lots of glue and silicone gel work wonders. Try to avoid unnecessary lengths of vacuum hose. Most hose is built to withstand internal pressure (inflating forces). Some hoses tend to collapse when the vacuum (very low inside pounds per square inch) is high. The shorter the hose, the better. Likewise, small-diameter, thick-walled, heavy-duty hose works best. I like to use short lengths of propane gas hose and valves. Screw-to-close valves, like those on propane tanks and hoses, work best. Be sure to cap the valve after the vacuum pump is disconnected.

Some persons have trouble sealing panels that abut. Again, generous applications of glue—I like hot transparent glue from an inexpensive glue gun—and generous portions of silicone gel do an excellent job. With a few exceptions, the things you glue together will not bear weight or other stress. The glue and silicone gel mostly serve to hold them in place and make them airtight. There are exceptions, such as attaching the handle on which you pull to close the inner chamber door. One of the many super-adhesive glues should be more than strong enough to bond handles to doors.

WHERE TO BUILD THE BUGPROOF (GLAS) ROOM

The location of your bugproof or bug-resistant room

depends on you and your professional needs. That is true whether you are a lawyer, a contract killer, a drug dealer, or the operations officer of a secret strike force.

The need for professional-quality safe rooms for conferences is more common than most people recognize. I know of one major hotel that rents its bugproof conference room for $250 an hour. (I've seen it, and it's not even bug-resistant, just impressively swept by a hotel employee in a uniform with a black box who then stands guard outside the door during the meeting.)

Whatever your motive—security, self-promotion, or capitalist greed—you must first estimate how important maintaining a low profile is for you personally and for your clients. For example, if you are a lawyer who specializes in acquisitions and mergers, your clients may not want to be seen together. If that is the case, be sure there is a logical reason for you and your clients to be seen in the vicinity of the bugproof room.

It is sometimes appropriate to locate your bug-resistant conference room away from your office and to fly a false flag over the entrance. For example, the raised letters on a brass plate on the solid outside door might read: "Zurich Custom Jewelry Importers—By Appointment Only." Nobody would think twice about special security locks and procedures at a place like that if they chanced to see you or a client enter or leave. A quick illustration from the world of espionage may be helpful.

Agents are almost always nationals of the country being spied on. Foreign case officers run the agents. Meetings between agents and case officers are touchy, especially since agents tend not to be experienced professionals and often are visibly nervous. For centuries, case officers have routinely met the agents they handled in whorehouses. Why? Because married men are nervous about their wives, bosses, security officers, or others finding out they use prostitutes. And men and women from all levels of business and society use safe-sex prostitutes.

You probably won't want to install a bugproof room in

Sally's Pleasure Palace, but a neutral location midway up a tall office building where your client might logically visit his bank, his accountant, his attorney, his barber, his broker, or his custom import jeweler will do nearly as well.

Real world bug-resistant and bugproof installations are not very exotic. In about 90 percent of the fixed locations with which I'm familiar, the bug-resistant conference room was an openly acknowledged and logical modification to an existing meeting room. If you upgrade a working conference room, you may have to break some hidden-agenda use patterns with a formal announcement: "For security reasons we have made a significant financial investment to make our conference room free from possible surveillance. As a result the conference room will no longer be available for casual use, lunch, or coffee breaks."

Controlling access to the room is essential. As you consider where to locate your room, how to limit and control access, and whether it really needs to be bug*proof*, you may want to review earlier chapters in this book. Whom you allow into the room and whom you tell about the room become almost as important as the room itself.

Most persons don't recognize they have a secret until it's been compromised in some way. If you've got a true secret—one that you alone know—you must clearly define *whom* you need to tell, *why* you will tell them, *when* you will tell them, *where* you will tell them, *what* (how much) you will tell them, and *how* you will tell them. The only true secret is one that you don't share with anyone and about which you *never make a written record*.

A final parting shot. Don't let anybody tell you their secret. If they don't tell you—and their secret is leaked— they will know you didn't tell. And vice-versa!

THE TROUBLE WITH TIME

It is always appropriate to plan security countermeasures in terms of the future. Today's associates and trusted friends may be tomorrow's competition or enemy. Lots of otherwise secure folks have gone to jail because a "friend" gave them up years later to save his (or her) own skin. Every living person is subject to bribery, to threats, and to many other forms of extortion. Over time, every situation will change, and every person in every situation will change.

Try, if you will, to imagine a current relationship that is so committed, so secure, and so filled with faith that it can't be tempted by some tomorrow's chance encounter, time's constantly changing spheres of influence and social contact, or the evolving physical, political, and financial circumstances in which we each must live.

"Ill luck, you know, seldom comes alone."

—Miguel De Cervantes, *Don Quixote,* (1605), Part I, Book III, Chapter 6, page 135.